Nature's Thumbprint

NATURE'S THUMBPRINT

*The New Genetics
of Personality*

by
Peter B. Neubauer, M.D.
and
Alexander Neubauer

Columbia University Press
New York

Columbia University Press
New York   Chichester, West Sussex

Morningside Edition with new preface
Copyright © 1996 Peter B. Neubauer, M.D., and Alexander
Neubauer
Copyright © 1990 Peter B. Neubauer, M.D., and Alexander
Neubauer

Library of Congress Cataloging-in-Publication Data
Neubauer, Peter B.
    Nature's thumbprint : the new genetics of personality / Peter S.
Neubauer and Alexander Neubauer.—Morningside ed., with new pref.
        p.   cm.
    Includes bibliographical references and index.
    ISBN 0–231–10441–3 (pbk. : alk. paper)
    1. Nature and nurture.   2. Genetic psychology.   3. Individual
differences.   I. Neubauer, Alexander.   II. Title.
BF341.N46   1996
155.7—dc20   95–37858

Casebound editions of Columbia University Press books are printed
on permanent and durable acid-free paper.

Printed in the United States of America

p 10 9 8 7 6 5 4 3 2 1

This book
is for Susie
and for Arthur

To find for each person his true character,
to differentiate him from all others,
means to know him.

<small>HERMAN HESSE</small>

Contents

Acknowledgments   xi
Preface to the Morningside Edition   xiii
Introduction   1

PART I

CHAPTER ONE
Nature, Nurture, Twins, and Others   15

CHAPTER TWO
The "Stranger" in Our Midst   25

CHAPTER THREE
Traits and Personality: Carving Nature at
the Joints   37

CHAPTER FOUR
Timetables of Change: Maturation   55

CHAPTER FIVE
Timetables of Change: Development   75

PART II

CHAPTER SIX
The Bridge Called Adaptation   93

CHAPTER SEVEN
The Vulnerable and the Invulnerable   113

CHAPTER EIGHT
The Individual "at Risk" in the
Environment   129

PART III

CHAPTER NINE
A Long History Briefly Told   141

CHAPTER TEN
Flexibility in an Ordered World   153

CHAPTER ELEVEN
*Nature, Nurture, and Implications for
Psychotherapy*   167

PART IV

CHAPTER TWELVE
*Living without Grandparents: The Loss of
Intergenerational Transmission*   183

CHAPTER THIRTEEN
*Individuality and Groups: A Final
Consideration of the Human Reliance
on Others*   195

Epilogue   201
Notes   205
Selected Bibliography   211
Index   217

Acknowledgments

We wish most of all to thank John L. Michel, our editor, and Pam Bernstein, of Pam Bernstein & Associates, our agent, for their enthusiasm and professionalism; Dr. Scott Barshack and Dr. Jonathan Stern for their invaluable research; Helen Fried for her technical assistance; Jane Isay for her great help and support; and Josh Neubauer, Edgar Rachlin, Ruth Birnkrant, Amanda Garland, and April Stevens for of course . . . everything.

Preface to the
Morningside Edition

S ince the first printing of this book, in 1990, waves of genetic research have been filling the news. Each week we learn more about how the structures and functions of the human body are genetically coded and how these "ground plans" affect not only the onset of certain diseases but also, most startlingly, human development and personality – the subject of this book. It has been a remarkable period, the results of which can hardly be overestimated.

In terms of discovery, for instance, the numbers are plain: Of the roughly seventy-five thousand genes in the human body, thirty-five thousand have now been identified and cataloged, compared with two thousand identified a mere five years ago. Along with this new mapping of the chromosomes comes the ability to decipher the roots of genetic diseases – from cancers to schizophrenia to Alzheimer's – while many psychological disorders once thought to be solely environmentally governed – such as bed-wetting – are now also linked to the gene. The idea of disrupting the course of these disorders may seem nearly unthinkable. But in a decade, becoming immunized to colon cancer by having a gene switched on or off might be something to do on a rainy Tuesday afternoon.

The past five years have also seen the sudden arrival of Prozac and its siblings as a means to combat depression and, ostensibly, to alter personality itself. The effects of these drugs point again to the strong, specific, and ongoing relationship between biology and behavior, and indeed doctors who dispense them might do better than to use them as generic, first-try ammunition against every emotional problem. Perhaps now more than ever we need to stay sensitive to the complexity of individual development. Those who seek to reduce this complexity by arguing for either nature or nurture – as in recent, well-publicized claims about the heritability of intelligence in groups of people – are truly missing the point. Constellations of DNA afford a wide array of possible expression and must be seen as *individually expressed,* from child to child and the adult each child becomes. We inherit ranges, as we will see, not blueprints for exact replicas.

What has not changed since 1990 is the way new information demands a new form of thinking on the part of parents, caregivers, educators, social workers, psychotherapists – in fact anyone interested in identity, one's own or others'. We hope this book will continue to serve as something of a way station between news flashes, a close-in view of heredity and environment as intimate participants in the unfolding of each human life.

P.B.N
A.N.
New York City
September 1995

Introduction

Genes determine eye color and blood type. This much is easy to accept. Genes regulate hormones, instill the sex drive, and contribute to the shape of our fingers and the length of our lives. As living beings, we share the biological roots that genes coordinate.

Strangely, though, these roots are not the ones we usually cherish.

Although each life begins with a combination of genes different from any other – making each of us, quite literally, a new experiment of nature – when we think about what makes any one person unique it is learning and experience that intrigue us the most. They mean much more to us, it seems, than chemistry and heredity. Who our teachers and parents were, what cities we visited, jobs we held, and interesting, provocative people we met, these and not biology are the life dimensions we tend to revere: they and not DNA are the stuff of literature, biography, and friendly conversation. Our grandparents may often have enough perspective on a family to be able to say, "Many of us McCoys from far back, on both sides, had perfect musical pitch," or "were troubled learners," or "were quick to anger." Grandparents know firsthand of reappearing traits in family bloodlines, the special hues and marks that characterize related members. But as we cast off the large extended families of the past, living instead in smaller nuclear units, in single-parent homes, or alone, the long-range view of families is easily lost – and *our environment* after birth is considered the only potent molder of our lives.

Who would have thought, then, that shyness may be genetically influenced? Or some eating disorders? Or a tendency to obsessive behavior, or an ever-widening range of physical and psychological illness? Who would have expected that one of the most humane of all human traits, empathy, had biological underpinnings?

Characteristics from ear infections to curiosity have come under intense focus by researchers in just this way. And more. As we will see, the very processes of growth, namely maturation and development, can now be seen to have inherited ground plans.

These findings have not been easy to accept. They challenge basic theories and at first seem to minimize the cherished power of the environment and our magic wish to mold our children according to our own hopes and intentions. Rather than being equally excited by the variations that genes initiate, we still tend to shy away from recognizing them. Rather than esteeming nature's generosity and variations, we fail to observe all the differences that make each life unique from the start. *We fail to see that individual makeup at birth colors each child's response to the world in which he grows.*

In so many ways, the most crucial ways, recognizing the role of both heredity and environment in the shaping of each person is not just an academic exercise; it is alive with implications for our day-to-day decisions. How we view the heredity/environment (nature/nurture) issue slants the very way we look at and behave toward our children and the adults they become, influencing teachers teaching, therapists analyzing, parents parenting, and poets musing. For this reason, it is important to understand what is in our overall cultural mood that resists information on the influence of inherited factors. The environment, almost as a rule of thumb these days, receives too many of the laurels for life's accomplishments and too much of the burden for its failures. And the fields of mental health, education, anthropology, and other allied professions have for nearly a century reflected our feelings. People hesitate before accepting the importance of genes in making us who we are, and we know some of the reasons why.

Issues such as slavery, racism, and Nazism molded our suspicion of heredity, and liberating social notions from Enlightenment philosophy to the Declaration of Independence to socialism have contributed to our faith in limitless change – both society's ability to change itself and men and women's ability to change themselves, to become not only what they want to be in life but also *who* they want to be. There are other influences as well.

Nearly fifty years ago, child psychologists, psychiatrists, and psychoanalysts initiated a new approach to human development. Departing from a focus on drives that prevailed in the 1920s and 1930s, theorists and researchers began to explore the individual's adaptation to the outside world, particularly the child's relationship with his parents. The success of this new approach influenced a growing awareness in many circles of the importance of social experience, how much it helps shape human development – and everyone's approach to child rearing changed because of it.

Fascinating studies of mother-infant interaction began to indicate that from the very first days of life children respond vigorously to stimulation from the environment and have the capacity (and need) to bond emotionally with the persons close to them. If this notion seems self-evident today it is only because the last decades have been rich in its exploration. The work of a number of researchers influenced everyone interested in children. Among them are Anna Freud, who addressed herself to the continual psychic changes that go on in child development, as well as to their place in psychoanalytic theory; Jean Piaget, who outlined the stages of cognitive development and how the child interacts with his world; John Bowlby, who evolved theories of attachment; Margaret Mahler, who carefully delineated the steps in the interaction between mother and child during the first three years of life; and others who detailed the existence of mental illnesses in childhood never before described.

As information became available about the significance of the human environment in the early life of the child, it was the role of the mother that seemed most crucial: her influence came to be seen as the essential ingredient for healthy development, and the consequences of her absence, either physically or emotionally, were considered truly devastating. Trauma (any external event that disrupts psychic function and development, such as a mother's death) was considered a major cause of most pathologies, and since mothers were assigned so much power and influence, most childhood problems were viewed as directly related to a defect in the mother-child relationship. Even conditions such as schizophrenia and autism, now known to have neurochemical bases, came to be interpreted as a reflection of the mother's inability to nurture her child. Theorists tied psychological disorders to the mother's attitudes, with various "types" of mothers causing different conditions in children, from psychosomatic illnesses to speech disorders to depression to antisocial behavior – in other words, the whole range of early pathology.

This purely environmental position is still with us today. An emphasis on the influence of parenting remains the trend, as though any diminution of this influence would limit the need for education and parenting, reducing the very power of our love and care. Learning has become accepted as the prime mover for what we are and will become, whereas genes are relegated to a different world, people tend to think, a world that is cellular and chemical and has little to do with *our* psychic world, with who we are as functioning, creative individuals. As a result, our genetic makeup, like a rainbow, becomes an elusive thing. Most people accept its role in the physical traits they see and for certain finite diseases like sickle cell anemia and Down's

Syndrome; but for various reasons they miss it in areas such as personality formation, where heredity and environment are difficult to tease apart.

It was in this atmosphere many years ago that an opportunity arose to follow the development of identical twins from infancy. With great curiosity a number of us decided to study the influence of the environment on the child, and by environment we meant, primarily, the mother's relationship with the infant.

Until then no identical twin study had explored twins reared apart as they matured and developed from birth on. Keeping pace with the child as he grows and passes through the varying phases of the life cycle is known as a prospective approach. All other identical twin studies were retrospective; they brought together mostly adult identical twins who had been reared apart, measured their similarities and differences, and tried to piece together the history of their lives. Yet such a retrospective approach had several drawbacks. Though some of the collected data were from school and work records, most of it relied on interviews and memories. People are not, as we all know, completely reliable informants about their lives and pasts, whether they are parents speaking about their children or adults about their childhood. Moreover, the first few years of life, so central to our interest because they are thought to mold the patterns of future behavior, are simply inaccessible to memory.

Most important, since the available data from retrospective identical twin studies more easily lend themselves to the limited investigation of illnesses and physical traits than to the examination of human development, our method provided a means of unraveling the intricate processes of growth itself. We could look at change *as it happened*. We would be there at birth and continue regular, intensive observations of separated twins and study their relationship with parents and siblings, collecting as much information as possible about behavior and growth. We made the assumption that once we understood development we would then be in a better position to understand the conflicts and disorders, such as phobias, that arise when a child deviates from it. In fact, it was our assumption that only by studying development as it happened could these disorders be accurately understood. Our study would therefore be useful to the investigation of both healthy and pathological growth, as well as to the ways the environment influences that growth.

The results of this longitudinal study, together with those from other observations of twins and infants plus a wide range of new findings in genetic research, have helped change our view of the relationship between nature and nurture in development. The broad implications of these results appear throughout these pages.

In general, what we and others found was that in the earliest observations of the first few weeks of life the similarities between separated twins reared apart were greater than we expected, and the differences in parenting, though always considerable and sometimes powerful, did not have the effect we imagined we'd find. Identical twins reared separately showed a likeness in the timing and pattern of development and maturation that was truly surprising, as well as a likeness in some of the foundations of temperament and behavior, from sensitivity to activity to emotional response. It appeared that an infant's unique way of engaging the environment and making it respond to his needs – plus the stability of these features throughout life – was influenced by genes and not only created by mother, father, or school. When a supportive, nurturing environment was available to the child, the true power of his inherited tendencies and susceptibilities revealed themselves.

At the same time, the field of genetics has made an extraordinary leap forward, from the discovery of the double helix and the admission of DNA into our regular vocabulary, to the results of genetic engineering that will soon – indeed have already – become a practical part of medical intervention. Almost daily new findings are published that confirm the pivotal role genes play in our lives. The gene for Alzheimer's disease, a mental crippler affecting no less than two million Americans each year, has been pinpointed on a small region of human chromosome number 21. One of perhaps a multiple of genes that cause manic-depressive disorder has been found near the tip of chromosome 11, Huntington's disease on chromosome 4. In time all fifty to one hundred thousand genes on each of the twenty-three pairs of human chromosomes may be identified, the very building blocks of life revealed micro-inch by micro-inch on maps of the chromosomes. This project is a spectacular undertaking, requiring both billions of dollars and dozens of years, and one whose consequences are truly mind-boggling.

As this research proceeds, the technology to transplant faulty genes with segments of new, healthy ones may be perfected, meaning that diseases as debilitating as schizophrenia, as fatal as sickle cell anemia, or as debilitating and then fatal as Huntington's disease or many of the other four thousand known inherited diseases (one-fourth of which affect mental functioning) might become, from the root, wiped clean. Gene banks already exist that accept deposits of human DNA so that family members can evaluate their inherited susceptibilities to various disorders. By looking at specific gene markers in one's own DNA and that of other relatives, diseases such as lung cancer may be traceable and, it is hoped, avoided. Since DNA can be frozen and stored, a person's great-great-grandchildren would be able to scan the branches of the family tree and establish their risk to life's illnesses. By knowing scientifically the actual diseases to which we are predisposed, we can then do everything in our power to avoid them, including – perhaps one day not so long from now – actual, individual genetic recombination.

Of course, the significance of nurturing isn't minimized by considering the role of genes. As we will see, the importance of providing appropriate care, the power of human interaction, and the positive effects of therapeutic intervention are all alive and visible. No one can curtail the strength of the environment, even in the face of a genetic matrix that predisposes growth. What we can do, however, is begin to focus our vision on the inherited patterns of nature as they unfold in development and influence our adaptation to life. What we must do is understand that we respond to the environment in ways biased by our innate makeup, and furthermore, that this makeup helps select the environment we respond to.

It becomes plainer by the day that the either/or choice of nature or nurture falls far short in explaining the complexity of human life. The polarization of the debate must be dismantled, and we can consider this alternative: a complementary series of influences where at one end environment fashions certain behaviors, and at the other heredity alone molds various traits and disorders – but in between, where most individual features reside, lies the complex interaction of both extremes.

In characteristics like physical growth, the biological side is certainly more observable; in others learning takes center stage. The continuum of influences varies from person to person and dimension to dimension, which is one reason why any answer resorting to numbers and ratios ("Genius is one percent inspiration and ninety-nine percent perspiration") is, Thomas Edison notwithstanding, just too simple. It reduces an ever-changing equation to a thing. We all want to know the answer to the question, Is it heredity or is it environment?; yet *what we must discover instead for each trait in each person is not a number but a specific, ongoing relationship.*

What we must begin to see is not the influence of nature

and nurture at a static moment in time and for a specific trait but the inherited patterns of maturation and development that unfold in the environment through the life span. People change; we must follow these changes.

And so, just as we can come to know, through introspection and therapy, the experiences that have made a lasting imprint on our lives, we can also begin to recognize the role of genes in the unfolding of personality. That knowledge, though more evasive, is no less real. And as it comes it will enrich our understanding of human life, not limit it, as the laws of heredity are often accused of doing. This new understanding is not a return to biological determinism in the old sense, or in any other sense. It is rather an attempt to broaden our vision of growth to include what shapes us from within as well as what shapes from without, and to look carefully at the exact nature of that interdependence.

This, then, is a book about accepting into one's life both hereditary history (nature) and environmental history (nurture) as interactive parts of human individuality. Since this is neither a research study nor a textbook, it is not our purpose here to make a presentation of identical twin data. We rather want to view the relationship of forces that shape our individual dispositions based on all the information available today, of which twin studies offer one revealing part.

And we emphasize the word *individual*. Although history has seen many theories about the genetics of groups of people, much of that guesswork either arose from prejudice or stimulated it anew. The subject of heredity has provided our recent past with so many examples of racial, national, or gender bias – leading to oppression or mass extermination or both – that its recounting is the subject of another book entirely. Unfortunately, there are still enough proponents of genetic determinism in our society (Shockley, for instance, on the question of intelligence; members of the KKK in the area of race; others in that of gender) that we must all stay alert against their claims, against the easy slide from specific genetic influences in the individual to the complete hereditary determination and supremacy of groups. Rather it is the fascinating and complex variations between growing individuals that concerns us in these pages – and we hope these are the particulars that mean the most in the reader's daily life.

Part I presents a background for this understanding, leading the way with case histories before diving headfirst into one of the most crucial topics – nature's ground plan for personality and growth. Part II describes the essential consequence of this ground plan, namely the process of adaptation throughout life, as well as two of nature's most interesting examples of adaptability. In Part III we look at some of the broader contexts of heredity and environment, including that of psychotherapy. And finally, in Part IV we focus on some of the implications of this view of nature and nurture, which demands most of all a new and vital awareness of the individual.

Since each chapter touches on a different issue, highlighting different problems or wrinkles, occasionally repeating earlier notions with each new twist, some chapters are more complex than others. To help illustrate the discussion and broaden the range of our examples, we have borrowed from cases in the twin and nontwin literature as well as from general and personal clinical experience. In certain chapters there is greater emphasis on childhood than in others, but the themes that apply to children of course apply as well to the adults children become – and are still becoming. The issues span the life cycle. Parents may find special connections as they see innate patterns in themselves and become sensitive to those that unfold in their children. But whereas parents and educators may be most interested in children, therapists may focus on adults, everyone on his parents, grandparents on everyone, and all of us on ourselves. We each have a similar task, which is to come to know who we are as biological beings continually responding to our environment in ways that are our own.

This book is about seeing our children, parents, siblings, and our own life histories as expressions of an ever-present, ever-changing relationship between genes and the environment that nurtures them.

Can observing, investigating, and wondering about genetic makeup, the "givens" we bring into the world that belong to our disposition, become part of the repertoire of self-study?

Can parents continue to accept the hereditary transmission of their children's physical traits but fail to accept the transmission of other qualities, such as sensitivity to noise, strength of self-control, even shyness?

Can we continue to believe that children are only who we make them and not also entities unto themselves, with biological pasts and developing futures?

If room enough is found, a comfortable niche, to accept rather than fear biology as an intimate partner of learning, we will see how nothing less than the potential for healthy growth is enhanced.

PART I

The living pattern latent in all growth.
GOETHE, *DAEMON*

CHAPTER ONE

Nature, Nurture, Twins, and Others

It is the common wonder of all men, how among so many millions of faces there should be none alike.

SIR THOMAS BROWNE, *RELIGIO MEDICI*

T he experience of living, as mythologist Joseph Campbell once suggested, is of greater value to human beings than the meaning of life.

Meaning is abstract; experience is real. Experience involves us all in the activity of the world – it puts us out in the arena of living where sensations, deeds, thoughts, and even solitude demand from us recognition and response. How we experience this world depends on that confederation of past events and present situations in which we live and grow.

Yet what we bring to the world also influences how we experience it. What we bring – our temperaments, patterns for growth, individual inclinations, and personal susceptibilities – affects the way we see life and the way we respond to it, from the first moments after birth to the day of our death. Simply put, what we bring is the gene.

Although the environment influences us unremittingly, we are also biological beings. As members of the species *Homo sapiens* we collectively share a range of ways to behave. And on top of this, as individuals, each of us has a set of genes different from those of anyone else and so has an individual range of ways to behave. The experience of living can change us only as far as our individual and biological ground plans for change allow.

Slowly we become aware of the forces that draw their signature across our living patterns and shape the way we grow and behave. And one of the best ways to become aware of these forces is to view the lives of people. Here are two:

A Responder

A thirty-year-old married man chose a profession that demands no social interaction. He is a highly skilled technician, dedicated to his work and to his precision instruments. His wife is from a different culture and speaks his language in only a limited way; she complains that he is distant, unsupportive, and unable to engage in a meaningful relationship with her.

In his earliest years he was reported to be a very calm, almost passionless baby referred to by his nurse as "Chinaman," implying that he was inscrutable, that he wouldn't let himself be easily "read" by anyone, that he hid his feelings.

Since no one knew about his fantasy life and the ferment of his internal psychic world, his parents always assumed he had a normal range of emotion. He never gave any evidence of internal conflict, for he was always quiet, and with his unassuming and unassertive demands on the adults around him he was thought to be a very well-behaved child. In fact, he was often praised for his reasonableness. Once, when he was just one year old, his mother put him on a sofa while she did chores around the house until she was suddenly called away to deal with a neighbor's emergency, forgetting her child was not in his crib. When she returned an hour later he was sitting where she had put him, quietly self-contained and seemingly content with the limited stimulation around him.

This man as a child often accepted being cuddled and comforted but on the other hand made no outward demand for it, nor ever initiated it. He was, in sum, a responder to life, never engaging others or showing signs of assertion to anyone at any time.

An Engager

At twenty-five-years old, this well-educated woman is actively searching for new experiences, is deeply committed to community affairs in her city, and is involved in various international programs of assistance for developing countries. She follows her own ideas, is action-oriented, and creates as well as enjoys sensitive relationships with the people around her.

As a baby she would bite during breast-feeding, revealing that the intensity of her actions toward the environment — her ability to take what she needed — was strong even on this early oral level.

As a young child she actively explored her surroundings, both visually and physically, and was so outgoing in her demands for contact that she intimidated even her mother. She imposed her needs in such a forceful way that her parents yielded to them rather than tried to control them, which would have been difficult in any case.

What is particularly interesting about this man and woman is that they are brother and sister. They grew up with each other, lived under one roof, and were reared by the same parents, but from birth on they revealed themselves as people with two essentially different ways of approaching the world. How we as individuals and how parents, educators, and psychotherapists view them largely depends on how we understand the relationship between nature and nurture.

Of the six billion base pairs of DNA in human chromosomes, unrelated strangers differ in only six million of them. The vast majority of our genes, then, are shared, identical in each person and representative of our species: we all have brains, hearts, eyes, livers, and limbs of a certain number and proportion, all of which work in a certain way for a certain time, age over a certain course and at a certain speed, and then cease to be. Our genes are largely responsible for all this. (In Chapter 10 we will look at this amazing order more closely.)

Nevertheless, each of us still possesses enough genetic

variation to make us distinct. Our skin tone, height, hair color, facial structure, posture, and blood type, for instance, are identifiably singular. Even brothers and sisters – conceived by egg and sperm from the same sources – differ in two million base pairs of DNA, and depending on the combination of their genes may appear strikingly similar to each other or astonishingly different.

People generally accept the gene's role in physical appearance, but they (we, all of us) tend to attribute everything else to experience. Individual personalities are most often understood as the inevitable outcome of people pursuing different lives, having different experiences, suffering "the slings and arrows of outrageous fortune," each in his own way. Any result can seem plausible.

Fortunately, however, there are new roads being built every day that lead us to a greater understanding of the genetic forces that interplay with the environment – and among these roads is a one-of-a-kind opportunity of nature, namely, two of a kind: two human beings with not just similar but the same genetic content. Although, as we've just said, brothers and sisters differ in only two million base pairs of DNA, identical twins differ in none at all. This is a biological fact with consequences, for as we study identical twins we are plainly confronted with some of the results of nature. What is striking about identical twins is exactly what we may overlook in the individual: how much of human distinctiveness, from physical proportion to emotional development to personality, rests on our genetic makeup. And this is true to a remarkable extent even when one twin has been raised apart from the other.

Experience modifies appearance, habits, life styles; social codes may be strict or flexible; environment may end the life of one twin at birth and nurture the other for a hundred years. Yet seen side by side, identical twins reared apart are a visible expression of more than this: they provide a view of shared biologies, which is why they are so interesting to study and why they contribute so much to our understanding of nature and nurture. They reveal in their similarities and differences what is more difficult to distinguish in the single person. Here are two examples, slightly modified from real cases.

> *Twin girls were separated in infancy and raised apart by different adoptive parents. Unlike fraternal twins, these girls were monozygotic; that is, identical, conceived from a single egg of the mother and sperm of the father. Each one was the other's genetic duplicate.*
>
> *When the twins were two and a half years old, the adoptive mother of the first girl was asked a variety of questions. Everything was fine with Shauna, she indicated, except for her eating habits. "The girl is impossible. Won't touch anything I give her. No mashed potatoes, no bananas. Nothing without cinnamon. Everything has to have cinnamon on it. I'm really at my wit's end with her about this. We fight at every meal. She wants cinnamon on everything!"*
>
> *In the house of the second twin, far away from the first, no eating problem was mentioned at all by the other mother. "Ellen eats well," she said, adding after a moment: "As a matter of fact, as long as I put cinnamon on her food she'll eat anything."*

And consider:

> *Identical twin men, now age thirty, were separated at birth and raised in different countries by their respective adoptive parents. Both kept their lives neat – neat to the point of pathology. Their clothes were preened, appointments met precisely on time, hands scrubbed regularly to a raw, red color. When the first was asked*

why he felt the need to be so clean, his answer was plain.

"My mother. When I was growing up she always kept the house perfectly ordered. She insisted on every little thing returned to its proper place, the clocks – we had dozens of clocks – each set to the same noonday chime. She insisted on this, you see. I learned from her. What else could I do?"

The man's identical twin, just as much a perfectionist with soap and water, explained his own behavior this way: "The reason is quite simple. I'm reacting to my mother, who was an absolute slob."

In the first case above, the adoptive mothers of the two girls felt differently from one another about their daughters' eating habits. One mother thought something terribly wrong with a girl who craved cinnamon on everything, who in fact made cinnamon a condition of eating. This mother was worried, and an obvious fight over food had begun. The other mother simply felt that her daughter was a perfect eater, which indeed she was if you discounted the cinnamon, as this mother did.

Despite different parental reactions, however, the twins maintained their craving. It could not be taught away, at least not yet. In a few years the taste for it might disappear on its own – we don't yet know; we only know that parental attitudes could not modify it at this time (though they probably affected the children's overall experience with foods they enjoyed). Had it not been for the fact that the children were identical twins, we might never suspect a predisposed taste for cinnamon – and we might then assume, as did the first mother, that her daughter was challenging her over the issue of food. We all know children who have different tastes at certain ages, often for sugar and sometimes for stranger flavors. If these preferences are genetically coded, then what at first seems to be the child's rebelliousness can now be seen as a natural predilection, which a parent may still insist on changing but whose origin is no longer either their own or their children's "fault." The underlying root is heredity. Often only the existence of an identical twin can reveal such endowed characteristics.

Similarly, maternal attitudes could not modify the men's compulsive cleanliness. Neither the mother who was compulsive nor the one who was slovenly made any difference in the eventual behavior of their sons. Both twins maintained their compulsion to cleanliness. Taken separately, we might accept either son's interpretation of his own behavior. Each man blamed his mother for the fact of his conflicts. Mother and mother alone must take the fall for this, they both said: she, the preacher of orderliness, or she, the slob. She did it! Without the awareness of an identical twin's similar pattern of behavior despite differing environments, it seems neither man considered the alternative cause – the one that lay within himself. Heredity is elusive. Most people have not yet learned to look for it.

Complex behaviors such as compulsions are not likely coded on a single gene – one, say, for spotlessness. We probably don't inherit scrubbing and disinfecting instincts. But cats do. And monkeys pick at each other's fleas. It may yet turn out that some complex behaviors, such as those that protect against infection, have been selected for in the course of human evolution, just as we have learned that certain mood disorders are linked to neurochemical and hormonal imbalances governed by genes. There are also other, more intricate ways genes influence certain disorders, as we will soon see.

Still, although identical twins share the same pattern of genes and everything is prepared for identical expression, this does not mean – and this is the key – that every trait will be identically expressed. What is endowed at birth is not a set of traits but a range of expression. The range is set by human evolution and the individual's inborn variations, and it accommodates flexibility. Our genetic programs allow for, and cannot thrive without, environmental influence. Heredity may even be invisible in certain situations. A trait for corpulence, to use an extreme example, will not be easy to spot if the predisposed overeater has only thin soup and salt-free crackers his whole, probably short life.

In the case of the men with obsessive tendencies, however, both developed into compulsive cleaners despite having mothers whose outward behavior regarding cleanliness was not at all similar. In the case of the "cinnamon girls," the taste for that particular spice expressed itself despite their mothers' different attitudes toward feeding.

An inherited predisposition to these characteristics was probably at work in both pairs. And we are learning more every day about the role of genes in our lives.

The study of identical twins reared apart is a fine tool that tells us much about the relationship between heredity and environment – the forces that in concert shape human characteristics and that, even more importantly, coordinate human growth itself. But it is not the only tool. Studying identical twins *reared together* offers different information, particularly about adaptation. We can view with interest the way each twin copes with the conditions of life, and since each has the other as a partner in development, we can see their interplay: the well-known "twinning" reaction, where one may become dominant, the other passive for a time, then reversing, each selecting or avoiding, understanding or denying the other. These reactions add much to our knowledge of environmental interaction. Another source of information is the study of single adopted children, where we can compare the ways they remain close to their biological parents, revealing the relative strength of heredity, and the ways they mold to their adoptive parents, revealing the influence of imitation, identification, and adaptation to the environment.

Combined with the ever-growing mapping and recombination of DNA itself, the study of these life situations discloses a wealth of knowledge about the interactive influence of nature and nurture. And what is revealed applies also to other life histories that at first glance seem almost too complex to decipher. These histories are our own. It is the challenge for each of us now to begin considering the forces that shape our lives and influence our very responses to the world. We have the tools at hand.

CHAPTER TWO

The "Stranger" in Our Midst

Thou shalt not oppress a stranger,
for ye know the heart of a stranger,
as ye were strangers in the land of Egypt.

EXODUS 23:9

N ot a minute into the world and we are already a part
of it. Doctors and nurses shower us with the care our
society expects, our maleness or femaleness elicits a
certain emotional response, and most of all we are given over
to excited and exhausted parents who have plans for us.

Parents welcome the newborn with all the signs that he
belongs – belongs, that is, to them. They consider his physical
needs, wonder who he is going to be, and imagine how he will
fulfill his role as part of their family (and culture) in both
physical form and emotional temperament. And yet consider
what so often happens:

> *Doris, an expectant mother, longed for the kind of*
> *cuddly baby boy her first son was not. Happily, her*
> *wishes were answered. The new infant seemed to share*
> *her disposition. She did not have to invent this child's*
> *love for her; the child brought it with him, and together*
> *they gratified each other's longings and created a har-*
> *monious, sometimes symbiotic relationship, loving,*
> *pleasurable, and deep. As time went on, friends got*
> *the impression that the two of them truly understood*
> *one another and, compared to the mother's relation-*
> *ship with her older child, preferred each other to any-*
> *one else.*
>
> *And what of this older child? From birth on he*
> *seemed more action-oriented, had a wider scan of the*
> *external world and a wider interest in the inanimate*
> *objects placed around him, such as toys. He did not*
> *especially like being held for long and never liked being*
> *cuddled; rather, he enjoyed exploring the world on his*
> *own. When the mother offered what she considered her*
> *greatest gift, the warmth of her love and affection, she*
> *received a disappointingly lukewarm response, and the*
> *more she imposed her love the more she encountered*

opposition. She began to feel rejected, and what's more, could not help feeling (unconsciously perhaps) that this was not "her" baby. Unable to be the mother she wanted to be – a mother on her terms – she was also unable to be one on his, and she blamed herself for the growing tension between them. The boy too began turning away from her and toward others who were more attuned to his own disposition, which led to even greater distancing from his mother and, in later years, to criticism that she never understood him, never cared.

The "stranger" in our midst, despite parental attachment and love, is the child outside our own way of experiencing and acting. *Our* children, it is felt, should fall within *our* sphere of understanding and satisfy *our* needs, their presence echoing our own. We have a harder time with the child who seems to bring to the world a variety of characteristics outside those imagined as part of our family history, one who in so many ways is his own person. Yet any child receives only half his genes from each parent; and as was mentioned in the first chapter, each child differs in two million base pairs of DNA from any of his siblings. Variation is not only possible, it is sometimes remarkable. The result may be a child who, at least at first, not only appears to be different from a given parent but in fact *is* different. Children come into the world as entities unto themselves. Yet becoming aware of the genetic forces that help configure the built-in variations of their lives has never been easy. We resist this notion partly because we simply don't look for it and partly because of the strong and natural expectations we have regarding our children.

What are these expectations?

A father expects a robust boy; he may get a gentle girl. A mother wants her child to have blond hair like her own; she finds it thick, dark, and curly. The parceling out of physical characteristics – the mouth is yours, the eyes and chin are mine – has always been part of parents' need to incorporate the child into the family image, the family identity. The fantasies of parents about a child – who they wish him to be – often remain quite uninfluenced by the child. This fantasizing may be a narcissistic wish on the parents' part, a desire to see themselves recreated in progeny, as God created man in His image. A certain measure of such wishing is inescapable – perhaps even desirable. It can be part of what binds parent and child together, a family myth or cultural identity that no individual differences are allowed to disrupt. Strong cultures survive by drawing new members quickly into their fold.

Yet although parents may influence their children and set before them ideal models for which they can strive, the parents can also interfere with the natural ground plans for growth that each person brings to the world.

> *In a case of identical twin boys reared apart one family battled against the studied, watchful passivity of their adopted son while the other family accepted this relaxed, unaggressive quality with relative ease. Although the temperaments of both boys remained steady through maturity and both ultimately chose professions that matched their natural inclinations (one accounting, the other academic medicine), the first boy's relationship with his unaccepting parents became embittered, and he has carried their disapproval with him for many years.*

Or consider this family:

> *The parents pride themselves on their literary, artistic, and intellectual interests. There are "music hours" at home when everyone listens quietly. The two oldest*

children, without apparent effort, sit unmoving during these home concerts with eyes closed and ears open.

 Yet the youngest child, a girl, is more activity-oriented than her siblings. She likes to run around, continually needing to be involved in doing and exploring things, which is seen by her parents as a sign of restlessness and hyperactivity. Action to them is viewed as a lower, more primitive form of expression compared to that of art and language. This child is indeed fortunate that she found teachers at school sympathetic to her needs. Even so, because she doesn't fit the family myth, its collective, self-imposed identity, she feels criticized, devalued, and just barely tolerated at home. In the minds of her parents there is no way she can become the student the older children are, because reading — something she does not yet love — is thought to be an essential instrument that cannot be adequately replaced by learning through action.

Who is this child? She is different from the others in her family and will remain so — not because her development is slow but because her mode of interacting with the world is different. It is easy for a parent to say she should catch up, as if her catching up would unite her with her family. It may not. If she indeed has a problem, special attention can surely help; but it will not necessarily change her in all the ways her parents might like or expect. What D. W. Winnicott says about twins applies as well to single children, that it is the parent's task "not to treat each child alike, but to treat each child as if that one were the only one."[1] This is a crucial theme, and it is echoed even more beautifully by poet Rainer Maria Rilke:

> Is it possible that one says "women", "children", "boys", not guessing (despite all one's culture, not

guessing) that these words have long since had no plural, but only countless singulars?

Understanding the singular child or adult means gaining insight into the process of becoming – insight into an individual's growth as well as his personality. As we will see in the next few chapters, some people by nature develop and mature more slowly than others. Or put better, in some growth is spread out over a longer period of time, as if in slow motion: even adolescence may be prolonged well into the late twenties or early thirties. Most children catch up eventually, but if continually measured against their peers their slower personal pace is often equated with immaturity, or worse, with below-average ability. Schools may reinforce this impression by suggesting that these children repeat a grade in order for them to level off with other children, but to be placed among younger students or among students with true learning problems may only send the wrong message to these children, that they have a problem in functioning.

Such obvious misreadings will continue until we recognize, using the landmarks of individual maturation and development, that rates of growth are a part of each child's makeup. Recognizing individual pace is a step toward accepting each person in his uniqueness, for nature provides a range of perfectly healthy differences even within the same family. It is the need to remake our children in our own or society's image that may push, pull, or bend them out of comfortable shapes – comfortable, that is, for who they are. Parents might be reminded that an infant's hair and eye color often change, muscles grow, fat disappears, timidity blossoms into self-assurance. The predisposed features of maturation and development are not stable or always predictable, and the inherited features one first feels are unexpectedly different may later seem perfect, those that seem perfect may later appear unusual. Although there may be conflicts that interfere with development, problems that do indeed demand special attention and appropriate intervention beyond those created by the parents' wish, the intolerance of differences is what tends to make any deviation especially troublesome – and it can happen as easily to a child who is precociously mature as to one who is too slow. For example:

> *Charles began to speak early, his developmental phases occurred early, and his interests were wider than those of his peers. Both teachers and parents were quite willing to assign to his precocious development positive values; they perceived him to be a talented and gifted boy. They accelerated him in school, making him skip two grades. But at present it is unclear how much this increased expectation has burdened him, for he has begun to appear to be a quiet, sometimes listless child.*

Perhaps the push to achievement was the right move in Charles's case – time will tell – but more often than not this push is made after considering only a limited period of growth. Parents who hope to strengthen family bonds by trying to fit their children into their own mold, traditional or otherwise, may or may not have luck, depending on whether the children themselves are inclined to that image of who they should be.

If a tall perch could be found from which we could view wide canvases – of our grandparents, ourselves, our siblings, and our children – we might see familial traits echoed down through the line, appearing, disappearing, and reappearing from time to time. We could see in broad brush strokes a pattern of continuity in physical and personality characteristics, and timetables of growth as well, that from a distance seem unbroken but from up close reveal, like a painting by Seurat, the distinctness of any given point. The points represent *us*;

they are our dispositions. Although we should be exceptionally cautious about typing an infant, assigning to him any unalterable features that may influence his emerging self-image, not to do so at all, not to observe nature in its unfolding can be just as misleading. Almost everything we do with our children is so highly charged with emotion and meaning that we need specific answers.

Fortunately, we can look to the child in his own right. Yet to see him *as he is* we have to learn to observe better, to recognize how he changes over time, to see development and maturation more clearly, and to know that each child has his own makeup. Each one relates to us in ways specific to his disposition, and with each we create a singular relationship.

There must be such powerful forces that limit our capacity to see, as if our own wishes restrict our ability to accept all the possible variations and differences in life. Perhaps in the future, rather than fearing these natural variations, we will become intrigued by them, looking to the freshness of creation with wonder and curiosity, and learning from our children as they learn from us.

This struggle is not new. It took thousands of years before civilization recognized children as children, before artists could paint them as they were and not as miniature adults, before people realized they could feel pain like adults and should be given anesthesia during surgery, before childhood schizophrenia and depression were diagnosed, and before the rights of the child were given a hearing and finally a manifesto by the United Nations. The Ten Commandments required only that the child honor his parents, never the other way around.

Given this history, it is not surprising that we still have trouble seeing the child for who he is. Of course, the child is part of us just as he is part himself. As the quotation that opened this chapter implies, we can know him as different because we were also different in our time, different from the people our parents expected us to be. By failing to recognize and adapt ourselves to our children's individual dispositions and rates of development, we not only lose sight of who they are, but we subsequently lose sight of how they in turn affect us. It is to the nature of the relationship between parent and child that we must finally look.

T. Berry Brazelton, who has studied infancy with so much insight, writes: "The evaluation of the behavior of the newborn infant has concerned many who are interested in understanding the relative contribution of the infant to the nature-nurture controversy. . . . Many researchers feel that *the individuality of the infant* may be a powerful influence in shaping the outcome of his relationship with his caretakers." (italics ours).[2] And Erik Erikson writes:

> Parents who are faced with the development of a number of children must constantly live up to the challenge. They must develop with them. . . . Babies control and bring up their families as much as they are controlled by them; in fact, we may say that the family brings up a baby by being brought up by him. Whatever reaction patterns are given biologically and whatever schedule is predetermined developmentally must be considered to be a series of *potentialities for changing patterns of mutual regulation.*[3] (italics in original)

By failing to recognize the child's own inborn timetables of growth, we not only fail to nurture in ways he most needs – ways specific to his own developmental patterns – but we also fail to see the ways he contributes to and regulates our own attitudes. We fail to see how he imagines and moves us. In short, we become blind to the dynamics of our relationship with him. "Ought we not to say that by fitting in with the infant's impulse the mother allows the baby the illusion that what is there is the thing created by the baby: as a result there is not only the physical experience of instinctual satisfaction, but also an emotional union. . . ."[4] This is Winnicott's way of expressing the notion that the baby "creates" his mother and gives to her all the power and magic she needs to satisfy him. It reveals that the mother is the provider, that she offers care, and that the baby in turn translates this attention into his own experience. What the infant makes of it is thus an illusion of what he wishes and needs her to be. And when he is dissatisfied, he will make of her a "bad" mother, with all the consequences.

Beyond the fantasies on both sides, the relationship between parent and child takes on a reality that it is our duty to understand.

We can be excited, curious. We can want to know all about this child because from the mix of the family's gene pool he is, at birth, quite truly a new combination of nature, as we were ourselves. He will follow his own path – though that path may be similar to our own – and the differences will enrich us. We have to learn to make adjustments to him, while he will use us as a model either to follow or from which to differentiate himself.

What parts will he accept, what parts disregard?

He will accept some of our fantasies about him and will evolve his own fantasies about himself, or better, about who he wishes to be. These wishes may change over time, demanding from us even greater participation in that change. This is an ongoing process, an intergenerational passing-on and reshaping of attitudes that offers the uncommon opportunity to test, challenge, and modify our outlook on family and self.

Such a position is not easy. As we have said, it runs counter to our narcissistic image of our children. In addition, it may silently confront tradition, that need of ours to maintain family and cultural values and preserve the family myth "as is." If tradition implies a pure repetition of the old ways, then any natural human variation will appear as a confrontation with tradition.

Yet there is another, more flexible definition of tradition that is poetically unmasked by composer Igor Stravinsky, who in the body of his life's work tried to remake old forms into new. He writes:

> A real tradition is not the relic of a past that is irretrievably gone; it is a living force that animates and informs the present. . . . Far from implying the repetition of what has been, tradition presupposes the reality of what endures. It appears as an heirloom, a heritage that one receives on condition of making it bear fruit before passing it on to one's descendants. . . . A tradition is carried forward in order to produce something new. Tradition thus assures the continuity of creation.

The acceptance of the child as a new, that is to say different, sometimes even strange, addition, does not violate our need for tradition. It enhances it.

CHAPTER THREE

Traits and Personality: Carving Nature at the Joints

I n the introduction to this book we wrote that the either/
or division of nature and nurture must be dissolved. Perhaps nowhere is this goal more important than in the
question of human personality – that constellation of characteristics thickly knotted with interaction, developing through
the life span, and bound in its expression to the society and
culture in which it emerges. Complex as it is, personality is not
unknowable, nor does it float here and there depending on the
wind of the moment. Personality takes a form, and the form it
takes has roots leading back to birth and even earlier.

Through new genetic studies, clinical observation, and
research on identical twins and adopted children, we are becoming increasingly aware that many of the human characteristics previously taken for granted as products of childhood
rearing and environment are rooted in the genetic matrix. How
and when these characteristics form in response to the life
experience of the individual is a question – one of the most
interesting and rewarding questions we can ask – that we must
always keep in mind.

Among the newest research on the genetics of personality is a twin study led by Thomas J. Bouchard at the University
of Minnesota.[1] After looking at hundreds of pairs of twins,
some identical, some fraternal, and putting them through a
battery of physical and psychological tests, Bouchard and his
co-workers found striking hereditary links to personality traits
such as leadership, vulnerability to stress, traditionalism, and
imagination. Other studies have found shyness to be exceptionally heritable. And still others look to obsessional behavior and
the susceptibility to addiction as genetically linked.

Time may well modify this list, and the methodology
of some studies, which rely too heavily on retrospective or
statistical approaches, might be questioned; but these as well
as other data (and the research and clinical observations noted
in this book) suggest that much of what we call personality has

an *underlying genetic foundation*. Although it was once thought that the traits of personality are shaped only by the environment over the course of development, we can now say that much of what becomes personality is intrinsic to the child from the start – in other words, predisposed by natural inclination. The constellation of characteristics that defines our temperaments is influenced by our genes because genes, as we will see, set the tone for the earliest ways in which we interact with our environment. Recent studies have found that even infants may reveal so-called Type A behavior: they are excitable, aggressive, and quick to anger even after they are taught how to relax.

This knowledge, that a child is genetically predisposed at birth to a quality such as excitability, is important, for we can then see whether such a quality remains stable through life or whether it changes, whether it is present all the time or depending on the situation, and whether there are other traits that can be expected to emerge or retreat into the shadows of character when circumstances change.

New studies are beginning to explain even this. One longitudinal program sponsored by the Institute on Aging followed ten thousand people over nine years. It found that some personality traits appear immutable over time. Friendliness and anxiety level were two of the characteristics that rarely seemed to change. Shyness, curiosity, engageability, and flexibility are other qualities that have been found to be durable. They offer us a glimpse into what we can call the *early building blocks of personality,* those that may become differentiated over time but which are observable even from the first few days of life. These are the threads of personality that weave their way through our lives even as our lives shift and change.

Of course, traits such as traditionalism, as studied by Bouchard's group, and engageability, as observed in infant studies, are qualitatively different from each other. The former is a composite of several personality characteristics often seen in adults, the latter a fundamental building block often seen in the newborn.

The critical question we must therefore ask in seeking to decode personality is one historian of science Karl Popper asked in a broader context, namely: how are we to carve nature at the joints – or, for our purposes, how are we to see personality traits not as we imagine them or describe them in adult terms but rather *as they really exist in early life*? The characteristics we identify must correspond, as Gordon Allport once said, to "the personal neural network as it exists in nature," that is, the actual hard wiring of the individual brain.[2]

Although it may be difficult to imagine personality as part of the genetically ordered network of the brain, it is not impossible, as we will see. These fundamental traits are not the only characteristics of personality, of course. Not all are so firm. There are also personal constructs, cognitive styles, reflections of social behavior, and the "discontinuities [that] are part of the genuine phenomena of personality."[3] But as we drift away from nature to include situational or social factors, we are also called back to it. "The fact that my age, sex, and social status help form my outlook on life," Allport also wrote, "does not change the fact that the outlook is a functioning part of me. . . . My tendencies lie within."[4]

And so, in order to understand whether there are overarching tones of mood to which people are initially inclined, principal and predictable ways they tend to behave and feel – agitated, enthused, wary, shy, sad – we have to look for the most basic joints of nature. Often they are hard to see:

A forty-two-year-old physician broke up with his girlfriend, with whom, in his view, he had been sharing a "meaningful" relationship. She accused him of drinking, of working too hard, and of seeing another woman on the side, all of which were true. It is a familiar pattern for him. He is an excellent achiever, esteemed by his colleagues and patients alike, and he follows advances in medicine very closely in order to have all the tools of his trade at hand to serve his patients. Yet at the same time he violates many rules, fails to pay his income taxes on time, has been cited for failure to pay parking tickets, and – though this is more self-directed a transgression than the others – visits his own doctor with little of the conscientiousness and care he applies to his patients. In fact, he almost never goes.

If it weren't for a close relationship with his sister, he would have none at all. He always felt abandoned by his mother, who involved herself intensely with her friends in the community and with charity work. His father was a caring man, an immigrant who had proved himself by building a successful business, but he too was mainly interested in his own achievements. Following his father, this man got A's all through his academic years. Recognition was found only in what he did, not in the sharing of feelings or shaping of relationships. Yet he was blind to this pattern. Ambition was part of his personality and organized his existence, only he could not recognize it.

The characteristics that are part of this man's personality are organized around a theme: ambition, achievement, and deception. They appear in the context of his work and his ability to have relationships with others, and they have been modeled on those of his parents. However, we cannot be satisfied with recognizing traits that have become differentiated and appear in specific adult contexts. We also want to ask about their antecedents, their foundations. An interpretation of highly defined adult traits that mentions only parental influences and situational contexts omits a crucial part, which is the underlying tendencies that *from infancy* shape how we meet the world.

Finding the precise gene for shrewdness, however, may well be an impossible task. Shrewdness is an adult characteristic that becomes specialized (that is, differentiated) over time, expressing itself in certain contexts. But perhaps perceptiveness and sensitivity are more fundamental versions of shrewdness, its precursors, and more easily linked to the genetic matrix. The task is to begin to see personality in its different forms, and the level we speak about must be appropriate to the specific population. Hospitality, which is an adult trait and expressed in certain contexts, is not an accurate carving of infant behavior. No one says, "Oh my, how hospitable your baby is," though "receptivity to touch" is an appropriate and useful way to describe an infant — it is fundamental since it lies close to the genetic matrix existing in nature. We have all seen infants who are more eager for interaction, more ready to receive cuddling, and more giving of their smiles than others. Conversely, an infant's predisposed high level of excitability cannot be linked to the behavior of an adult who, with layers of psychic defenses, may screen this excitability so that it emerges only at certain times or not at all — or even as something quite opposite.

In essence, then, the characteristics that are representative of a six-day-old girl may not be expressed in the same way in a sixty-year-old woman. And carving the basic traits has historically been a problem. Just look:

"Justice, verity, temp'rance, stableness, / Bounty, perseverance, mercy, lowliness, / Devotion, patience, courage, fortitude" are the "king-becoming graces" listed by Malcolm in act IV, scene 3 of Shakespeare's *Macbeth* (and poor Malcolm said he didn't possess any of them). Hippocrates suggested four distinct temperaments (phlegmatic, melancholic, sanguine, and choleric) as major divisions; Wilhelm Wundt, the famous experimental psychologist of Leipzig, designated as basic traits slow, quick, weak, and strong (and a person could fall in between any two), whereas for Carl Jung extraversion and introversion were the reigning personality categories, with "active" and "passive" as adjectives. (Jung writes, "One should speak of an *active* extraversion when deliberately willed, and of a *passive* extraversion when the object compels it" [italics in original].)[5]

These are but a select few of the hundreds of systems created throughout history to pinpoint people's basic temperaments. Horoscopes and tarot cards do similar things using their own, let us say, nontraditional methods; the *Diagnostic and Statistical Manual* (fourth edition, revised) is psychiatry's way of classifying mental illnesses and personality disorders; the Minnesota Multiphasic Personality Inventory is one of psychology's elaborate questionnaire methods for categorizing traits, using hundreds of statistically tested questions; there are measures such the Neonatal Behavior Assessment Scale devised by T. Berry Brazelton, in which a wide range of activities are accounted for, many related to perception – light, noise, touch – and others to human interaction – cuddling, consolability, and so on; and then there are projective instruments like the Rorschach test and the Thematic Apperception Test. If we add to this list all the nonsystematic, mostly unconscious approaches each of us uses to make sense of the people we meet, the number of individual systems swells to the number of individuals on the planet, and we would need a system to classify the systems.

It does not take long to see how limited Hippocrates was in his approach to personality types. With only four classifications and no "in-betweens," the variety of personality he could describe was abruptly cut short. But between Hippocrates' four temperaments and the confusing whirlwind of each person judging for himself the personality of others, some communal way of understanding and describing fundamental trends can be found.

What's in a name? In this case, everything, because in order to apply any of these characteristics correctly – and for connections to genetic makeup and neural activity to be understood – their keys must be discovered. The job has already begun. For instance, in a study done in 1989 of single (nontwin) infants, researchers following Stella Chess's work found clusters of traits that formed stable patterns over the first six months of life.[6] These trait clusters they labeled "activity, arousability, and intensity" and "approach, adaptability, rhythmicity, and destructibility." The first set were considered "drive" traits, those that activate the infant, whereas the second were "regulatory" traits, those that mediate the drives. The research team found that infants with a preponderance of the traits in one set had fewer of the traits in the other. But what was most interesting was that from six weeks to six months of age the trait clusters exhibited *remained fairly constant.*

This discovery of consistency in such early patterns of behavior is a formidable result: it speaks to a regularity that seems less imposed or conditioned from the outside than brought about from within. It enhances the notion of inner, intrinsic tendencies that over time, despite fluctuation, resolve themselves in a basic constellation that can be called *personality* – traits organized around a theme. And early predisposed influences at birth affect its adult shape, even if the final differentiated characteristics (taking names such as "hospitality" or "shrewdness") inevitably vary.

What are some other of these fundamental characteristics?

In our group's study of identical twins reared apart one outstanding feature was the early social behavior of infants. Some children had a limited capacity to engage the environment, stimulate the mother to a specific response, or impose their wishes and needs in such a way as to be heard. Other children, however, expressed their needs perfectly well and got rather quick satisfaction. The first group depended on parents to "read them" and satisfy their needs; the second group brought to the world a greater ability to demand that satisfaction and stimulation. These qualities might normally be considered a product of environment, of parents who elicit or hinder the activity of their infants. But since identical twins were available for study, we could see that when there was either a low or a high level of engageability in one infant, it *was shared* by the identical twin reared apart.

The ability to engage one's environment thus appears to be predisposed. We may be born with a potential for engageability that is triggered, to a greater or lesser degree, by offerings from the environment. It will also affect the adults these infants become. From clinical experience we know that some adults confront the world through action – by doing – whereas others prefer to formulate their thoughts and plans in a way that minimizes their need for concrete activity. Although this feature can be transformed and appear even as quite the opposite quality in adulthood, it may in its fundamental form exert influences that play a significant part in the individual's choice of profession, in his social interactions, and in the very pattern of his life.

Another telling feature of personality arises from the study of emotion, known also as affect. Not surprisingly, the child's affective state was found to be related to that of his parents. Children who were accepted as having problems and difficulties, who were received with tenderness and responded to with care, seemed to have a more positive mood than their twins with mothers who were critical, negative, and disturbed about their children. There appears to be a striking transmission of mood to the young child even though other genetic inclinations, toward illness for instance, remain the same. For example:

> *Identical twin boys had a tendency to alcoholism, but one twin, with a positive mood seemingly transferred from his parents, appeared to have much less reason than his twin to continue drinking, and indeed he stopped.*
>
> *In another case both twins revealed a history of gastrointestinal difficulty, though one twin, whose parents supported him, seemed less affected by it than his co-twin. This created an unusual situation. The happier twin always appeared healthier than his sadder counterpart in spite of the fact that they shared the identical problem.*

The mood that is conveyed to a child often masks underlying difficulties he may have, and as long as he appears happy the problem or illness may often seem insignificant, though in fact it is not.

Yet although children's moods are conditioned by parents, and these moods in turn change how the parents interact with them, emotion is also connected with specific regions of the brain and may have strong genetic roots. For instance, the gene markers for manic depressive disorders, as we have mentioned, have recently been located on the human chromosome. In spite of the fact that multiple rather than single genes may well be found to be responsible for this disorder, we can feel comfortable saying that depression is not necessarily the result of poor parenting; it is neurological and genetic. Other disorders in the way people express or receive emotional signals (autism, for instance) are also inherited. It is always worth considering a genetic and intrauterine influence in emotion as well as an environmental one.

Of course, emotion is a particularly interesting and difficult subject because it is most usually cued by outside stimuli. When one person makes another angry, it is a delicate task to distinguish between the situation that prompted the outburst and any anger residing in the person at the start. The same can be said, more strongly in fact, for a characteristic such as empathy. This most human of traits, that of relating to other people's experiences because we know how those experiences feel ourselves, has often been believed to be conditioned by the environment after birth. It is a comforting thought, that we *learn* to be compassionate and empathetic people. But is this notion true? Does it carve nature at its genuine joints?

Recently, researchers have sought to explain the direct link between traits such as empathy and the genes that govern brain activity. The ability to "read" faces and imitate them is present in primates and in human beings from as early as the first few days of life. A neurological system of transmission seems to exist from infancy, allowing for the information of social signals – smiles, frowns, cries, and laughter – to be received, understood, and re-communicated. Identification with these emotions is the next step, one which begins and matures predictably for everyone except those who have neurological defects in these emotional pathways.

Empathy is a challenging concept to think of in genetic terms. One researcher, Leslie Brothers, makes it easier when she indicates that what may be called empathy among human beings is the same as "social-emotional communication (in animals)" and "social signal processing (for neurons)."[7] They are all related to a single process. The many seemingly "adult" personality traits, although connected in significant ways to the environment that molds them, are at a fundamental level linked to more basic characteristics. Now find the root from which they spring – see shyness as "low arousability" or empathy as "social-emotional communication" – and suddenly they may not be so difficult to conceptualize: they are a functioning part of each individual's constitution at birth and a part of the adult personality matrix at maturity. The threads weave together throughout life.

A final fundamental characteristic that emerged in our study of identical twins appears in the way some infants appeared more human-oriented than others, following with intensity and frequency the faces and voices of people rather than the inanimate objects placed around them, like hanging mobiles and stuffed animals. Other infants had a tilt toward the inanimate world, and when offered a choice between the dangling key or the human face were more willing to fix on the key.

Strangely enough, this orientation was often correlated with birth weight. Lower-weight twins focused more on the human presence, whereas the heavier ones interacted with a wider range of stimuli, including inanimate objects. One possible explanation for this phenomenon, though we should stress it is tentative, is that from an evolutionary standpoint it might be beneficial for a weaker infant to focus on people, who can provide the food and nurturing objects cannot. Writes Donald J. Cohen, "The impression that the less well-endowed twin becomes more socially advanced may result from observations of his close tie to his mother."[8] Conversely, the twin with reserve strength has more freedom and opportunity for wider exploration of his environment – objects included – without risking his own well-being. In any case, this difference in orientation lasts at least over the first five years of life and often longer.

Since human versus object orientation was correlated with birth weight, one might be curious about the influence of the womb environment on various features of disposition.

Birth weight is influenced by factors within the mother's womb during prenatal development. It depends on the position of the fetus and the access to blood supply and nutrition. Brain laterality is also influenced by conditions within the womb, and it is the dominance of the left or the right hemisphere of the brain that in turn determines left- or right-handedness and many other hard-wired traits. Because of this intrauterine influence, then, it is possible that *some identical twins have non-identical features.* One twin may be left-handed and the other right-handed, one heavier and the other lighter, one human- and the other object-oriented, and so on. Although these features may be considered environmental in origin, they are not parental, at least not in a way that parents can do very much about them. A child's position in the womb is probably not genetically fashioned, but it also has nothing to do with child rearing.

This result provides one reason for caution in deciding on either nature or nurture as an early, final answer regarding the origin of personality. We must pay greater attention to prebirth uterine conditions that appear to have a hand in molding disposition. Without separately reared identical twins available for study, a clear distinction in this matter would have been difficult to make.

It has become clearer by the day that there are basic, genetically coded features of human personality that emerge at birth and influence the way a child orients himself to the world. What we have to be careful not to do is conclude that all adult personality traits must necessarily have a direct genetic source. We have just considered one reason to hesitate: conditions in the mother's womb may help fashion characteristics that look the same as inherited ones.

There are, however, other reasons for caution, among the most powerful and important of which is our slow, unavoidable adaptation to our environment during the course of living. We will look at both development and adaptation itself in the next few chapters, but here we might say this much at least. Pressures from society and culture and parenting may have effects no biologist can predict. This is not to say that our biologies don't accept or deny those pressures in their own naturally determined ways – only that we should be sensitive enough to the power of the environment to understand how it influences our own predisposed patterns. In discussing the reasons for caution when approaching human personality, we can even take a step back to imagine both influences, nature and nurture, at work together.

Compare personality to the water's surface in a wide bay. It can be seen to swell and circle, to surge in some areas and retreat in others. If our vantage point is a cliff overlooking the bay, on the one hand it is easy to spot the swells and eddies of the water. This is what people have tried to do in identifying personality patterns, attempted to say, "A person's character can be categorized according to particular traits that we can label," such as a water swell or eddy, like curiosity or shyness.

On the other hand, sadness for some can come and go in the course of a day and at certain times in life appear to increase greatly or to lessen. If a trait is not stable, perhaps there is a timetable for its emergence and disappearance that is genetically influenced; or perhaps it is not so strongly genetically rooted in the first place. Perhaps it is more dependent on an outside stimulus for expression. This brings up the question of environment in our model.

When a light wind blows up and creates a frosty shimmer over the water's surface, the underlying currents may be hard to detect. And when a gale hits, the surface water itself may change direction. Weather of one kind or another, like the environment in a child's development, is ever present. There is wind, rain, cold or sunshine making contact with the water's surface all the time. Does this outside weather determine the water's flow? It certainly makes a difference. Trying to piece together the influences of parenting, schooling, society, and culture on the characteristics of personality has been the long-sought goal of researchers in psychology, anthropology, education, psychiatry, psychoanalysis, biography, literature, and in a different way, visual art, too. We know that lack of care has a definite effect on development, as does the loss of a parent; we know much about children growing up in the midst of war, and the effects of trauma; we know something about the psychology of groups, how people sway, how they rebel; we know something of society's role in shaping behavior, and on and on. In addition, we know ways to solace, soothe, and intervene when the water becomes too rough. ("Why does pouring oil on the sea make it clear and calm?" Plutarch asked. "Is it that the winds, slipping the smooth oil, have no force, nor cause any waves?") We know that some effects of weather are profound, others weak, and that these effects have to be understood, for never is there no weather.

At the same time, the wind does not act on water that is still. We know that the ocean is moving and takes its shape from the contours of the ocean floor beneath. This ocean floor in our model represents the patterns of inherited coding, the genes, that incline the body in different ways. The ocean floor is invisible from our high vantage point above the bay, yet we know it is there. Even on a calm day the water still swirls, sometimes imperceptibly. The ocean floor has mounds, rifts, and flatbeds. *To be content with studying only the profound effects of the environment, the weather, is to miss the underlying forces of nature beneath, the gene.*

One gene, like a pebble on the ocean floor, may not in itself have a noticeable effect on the water above, its own particular ripple on the surface, so that we can say with certainty, "Yes, underneath that ripple must be a pebble," underneath the trait shyness must be a single gene. The connection between a gene and a surface trait is rarely one to one. There is probably no lone gene responsible for most of the character-istics we consider to comprise personality, a gene for curiosity, sadness, or fortitude.

There are, however, specific single genes for many in-herited diseases, such as alcoholism. And a cluster of such genes, like a mass of rock, may have noticeable effects on a variety of surface traits. We've been seeing that genes form certain fundamental characteristics, discernible unique patterns that at birth orient the body to the environment in specific directions, such as engageability and curiosity. Part of this pat-tern consists of specific ranges of flexibility that allow person-ality a certain latitude in expression. Genes code for these ranges of flexibility as well. Through their study we come to know the workings of the genetic matrix in the body – how it responds to life's varying conditions over time. And through the study of development, we begin to see the interaction of endowment and environment.

> *In 1941 a case was reported of two Japanese twin brothers who were separated at birth and reared apart. Unbeknownst to each other, as young adults they both suffered bouts of tuberculosis (perhaps not so unusual for men of this age) and stammering (which* was *un-usual). More unusual still were their divergent paths at the same time in their lives: one became a thief, got caught, and went to prison; the other entered the Chris-tian clergy.*

What should be made of such distinctive choices, especially in a society and culture that esteems conformity? These men shared the same country at war. They also shared the experience of adoption. Yet others who share such experiences don't al-ways strike out in unusual ways. These men, however, also possessed an inner personality described by the researcher who studied them as "weak-willed." They seemed naturally inclined to be swayed by two dissimilar but powerful environmental forces – crime and Christianity. Neither pressure, the internal nor the external, can ever be ignored.

Experiences and opportunities vary, as do the basic inherited variations of temperament. To get a better understanding of evolving personality, then, it must be seen in the context of the individual's adaptation to the world. In this long view of life, every change the genes stimulate is also tempered by a changing environment. The genes send out new signals, make new proteins, enzymes, and hormones, modify previous structures and functions, and trigger possible ranges of action. The environment influences this process, passively or actively, less or more, every step of the way.

There is no room in this dynamic, whirling house for a static, inflexible image of ourselves or our children, which is the reason for the next step: the necessity not just to see individual personality complete at a moment's time but rather as creating itself and unfolding throughout the life span – in the fascinating processes of development and maturation.

CHAPTER FOUR

Timetables of Change: Maturation

You may expel Nature with a pitchfork, sir, but she will always return.

P. G. WODEHOUSE, *VERY GOOD, JEEVES*

An article by Ira Berkow in the *New York Times* of March 14, 1988, quoted Roy E. Parke, director emeritus of the Strawberry Festival and Hillsborough County Fair of Plant City, Florida, on the subject of his precious strawberries:

> You don't want them to get buggy. If they do, you've gotta spray 'em with fungicide, of course. That'll help any strawberry. . . . But a strawberry has a responsibility of its own. It's independent; it's a part of nature. The flavor, the size, the color. . . . All the farmer can do is try to help it along. Give it a little potash, a little magnesium, a little water.

There is much you can do, in other words, much you absolutely must do to help life grow; but nature has its own path, too. A strawberry, not alone among living things, "has a responsibility of its own," a responsibility not in any moral sense but in its own necessary internal ground plan for living and growing. Consider a young girl whose accelerated growth makes her appear mature at nine years of age. One can practically see in her face and serious demeanor how she will look when she is forty – set, determined, and self-assured. Her puberty has come early, and with it a well-anchored identity. Compare this child to a boy whose maturation is slow, whose body is amorphous until late adolescence, and who cannot find his place socially among his peers until his late twenties.

The rates of physical and emotional growth in both of these children are linked not only to their life experience (which surely plays a part) but also to their inborn ground plans for maturation. Boys and girls, as everyone knows, mature at different rates in general, and this boy and this girl have individual rates that differ even more.

Such variation should not be surprising. People no more

inherit bodies than they do minds. What we inherit, to be precise, is a set of genes. All subsequent growth, both physical and psychological, emerges from one starting place: a fertilized egg and twenty-three pairs of chromosomes, each containing strands of double-helix-shaped DNA. The guideline for growth begins prenatally, accounting for the precise position and function of two eyes, a nose and a mouth, two arms, two legs, and a spine. And this guideline continues all through our lives.

It was once a common belief that genetic characteristics appear strongest just after birth, the pristine time before experience begins muddying innate behavior, covering it with learned responses. It almost seems logical that what the newborn reveals is his heredity, whereas in every year after his birth one sees only the additions created by his rearing and learning. We have been taught that the major changes in our lives are those determined by our life experiences – decisions, traumas, strange encounters, powerful teachers, and so forth.

But as has probably become clear by now, modern research suggests that this notion is glaringly wrong. We are not equal to the sum of our experiences. Rather, our inherited makeup (our endowment) converses steadily with experience and learning, not merely at birth but through the whole of life. This critically important wrinkle in the nature of individuality – on who we are – is often lost or forgotten. The Deists (Thomas Jefferson was one) believed that God created the universe like a clockwinder: He wound it up and let it go, leaving mankind to its own devices, sink or swim. But the genetic code does not take such a hands-off approach. Genes keep sending messages, keep switching on and off, whether simply to oversee the running of an organ like the heart, to signal a disease like Huntington's, to adjust our bodies during certain times in the life span as in puberty or aging, or to orient us to the world in particular ways. Genes influence us up to the moment of our death. Our programs are ones of change, and so our very growth depends on them – with some vivid consequences, described, for example, by Susan L. Farber:

> "George and Millan were separated immediately after birth and adopted by different families. George was adopted by a newspaper "makeup" man and his wife and was raised as an only child. The family moved around the country, settling in New York at approximately the time George began first grade. He did not know he was adopted, much less that he was a twin, until he was eighteen years old.
>
> "Millan went to a young couple in Salt Lake City and had a younger sister, also adopted. He completed four years of high school but decided against college. He was interested in music while George's profession was commercial art. Shortly before the age of eighteen, Millan learned he had a twin, and the two boys met at age nineteen for the first time. . . . They were handsome, athletic young men, so similar that neighbors mistook them for one another. Both had won boxing championships. Both showed indications of artistic interest. Physical characteristics were identical or almost so, down to the location of cavities in the teeth. They impressed the interviewers as markedly similar in personality, and tests of personality were so alike that one of them was rechecked because it was thought a mistake had been made (it had not). . . .
>
> "After a few weeks, the boys parted and did not meet again. George became a glider pilot and Millan joined the Coast Guard. Within a few months of each other in 1943, each developed signs of anklosing spondylitis (a progressive crippling disease of the spine). The

> *atypical progression of the disease was the same for both men, and each responded similarly to treatment.*"[1]

From athletic maturity and skill to the onset of disease later in life, these twins reveal remarkable parallels. Maturation is linked to our genetic makeup, the natural "givens" with which we are born. The broad outline of maturational sequences, whether they appear as distinct units or overlap, follows a rough hereditary guide in each person, one that asserts itself within a range of conditions. Changes such as the ones experienced by George and Millan reflect the fact that growth does not proceed in a seamless straight line from birth to maturity. Discontinuities mark its way, and in spurts of physical maturation and disease there are dramatic programmed changes in the way the body is organized, changes such as the individual taking his first steps, speaking his first words, and making his first sexual response. The ego apparatus of perception, cognition, language, and motor functioning imbues us with features that determine who we are and how we differ from the rest of the world, not only now but in the guided changes of tomorrow. Let us take a closer look.

Maturation in Childhood

In daily use the terms *maturation* and *development* can be heard almost interchangeably, and what is implied by using either is that a child is growing, aging, and becoming adult. There is an important distinction between them, however, and the dictionary reflects it. *Develop* means "to unfold; to evolve the possibilities of," whereas *mature* means "to be brought by natural process to completeness of growth."

The difference? Development has more to do with environmental influences on growth and change, including emotional and psychological changes, whereas maturation encompasses more biological, hereditary plans. Although this distinction may give us a good sense of how the categories can be viewed, it still leaves too much unsaid, and so let us take them one at a time, beginning with maturation here and saving for the next chapter the "evolving possibilities" of development.

As we have said, some effects of genetic timing are easily recognizable, especially those that shape the body, determine its changes, and incline people to specific hard-wired illnesses linked directly to the genes, such as sickle cell anemia (expressed in the first few years of life), Huntington's disease (in middle age), and Alzheimer's disease (in late adulthood). These timed changes fall into the category of maturation rather than development, for they are initiated by specific hard-wired changes in the body that are in turn spurred by our genes. But other normal maturational changes also emerge primarily from nature's genetic plan of growth, namely the motoric, perceptual, and cognitive changes that influence how we interact with the world in so many ways.

In childhood, think of what happens when our brains reach new levels of complexity:

At four or five weeks of age infants begin to smile socially. The earlier smiles are typically in response to internal pleasure, warmth, cuddling, physical satisfaction, and so forth. Now the very smiles of others – without differentiation between parents and anyone else – provoke a reaction known as the social smile. What could seem more environmentally determined than this, a baby's response to loving attention? Yet we also have to consider the genetically influenced maturation of the brain, which sets the timing of this response. One can observe that unless a child has been emotionally deprived or physically abused, identical twins reared apart and exposed to comparably nourishing families will both begin to make this response – a smile for a smile – within days of each other.

At seven months, nearly every healthy infant experiences a cognitive reorganization known as the stranger reaction: the sight of approaching strangers suddenly elicits a negative response from the infant, who could not before reconcile the faces of unknown persons with those of the known. This sudden reaction is also a predictable change, a milestone in the way the infant responds to the outside world, and one can imagine its adaptive purpose (although there may be others), which is to reinforce the child's primary attachments to mother and father – at least until the time comes when new faces can be tolerated and included in his ever-expanding inner and outer world.

Later, *at seventeen months*, there is another predictable shift in the way the world is experienced, a reorganization from a purely perceptual impression of images to an understanding based on symbolic and linguistic experience as well – in other words, from sensations to the ordering of language. Although this maturational milestone, which is important for all later development, depends on environmental stimulation for its expression, the underlying guide for its course and timing appears to be biological.

Jump to *two years* after birth and the effort involved in gaining bowel control. Learning is involved, needless to say – the wheres, whens, dos and don'ts of toilet training. Genes, however, have a steady influence on the maturation of the central nervous system and on muscle control, which in turn have the first say about when toilet training is biologically possible. The training has to occur in concert with the demands, abilities and timing of the child's body, and no amount of pleading or scolding, no matter how sincere, will work without it. In our group's study a delay or difficulty in one twin's toilet training was almost always accompanied by a similar delay in his co-twin.

A genetic timetable – some call it a genetic clock – is

partly responsible for stimulating the normal maturational processes like speaking, reading, and walking, and before moving on to the adult changes set in motion by genes, we can consider what actually changes in these early ways we experience the world.

If you ask yourself the question, "Who was I as an eight-year-old (a four-year-old, a two-year-old)?" you may recall the special memories of childhood: the people (parents, siblings, aunts, nurses), the places (homes, particular rooms, the color of wallpaper), the events (birthdays, magical trips to relatives or foreign countries, spankings), and clear moments of closeness, fear, and joy. Yet before considering such montages of memory that belong uniquely to each of us, consider the way you experienced each object or occurrence.

Although we place private importance on the scrapbooks and mementos of each of our lives, the experience of these objects, people, and events are filtered. How an infant sees, how he crawls, how he speaks, and how easily he does these things are the qualities that reflect the maturational process of his growth, the ones most clearly linked to the genetic matrix. In psychoanalytic language these qualities are part of the ego apparatus. They are the instruments through which the world is perceived and screened and with which we act, influencing our view of the environment and of ourselves.

During infancy, for instance, we react to stimulation both internally and externally. We react to sight, sound, smell, taste and touch (our perceptions); we exhibit a particular degree and quality of vigor in our movements (our sensorimotor skills); we exhibit curiosity and make contact with human or inanimate objects (our orientation, awareness, and response); and we exhibit an ability to understand, to think, and to speak (our cognition). These qualities may be slow or accelerated, responsive or unresponsive to stimulation, and they are measurable; taken together they begin to give the careful observer a distinct feeling about his own personal past or that of a particular child. Adult personality may not always be reflected in these features, yet much of what makes a person unique *will* be reflected. What accounts for individual variety stems from a whole set of these inner maturational faculties, the earliest forms of which, from birth, redefine our very approach to the world.

These dimensions are connected directly to known areas of the brain and the central nervous system, such as the centers for language, memory, and the vast network of nerves and synapses throughout the body. The coordinates of these brain loci and the functions they perform have long been studied – which is not to say that we know all or even much of what there is to know about them, but only that their existence is no longer controversial: we know that we inherit the genes that guide their growth.

These abilities, or call them faculties, are linked to the biological matrix, and the range of their possible expression is preset. Although the nature and extent of environmental stimulation modify what we see, how we move, and what we turn toward and away from, the internal faculties of maturation are not random. Regardless of the environment that awaits the infant in the world he will have a way of orienting himself to it that affects what he reacts to, what kind of representational inner world he creates for himself, and how he responds to people and thus how people respond to him.

The results of identical twin studies, including our own, reveal that many of these early faculties of maturation were concordant, that is, shared by an identical twin reared apart. In one case, both identical twins revealed a limited ability in verbal formulation and articulation. Speech came slowly, and at its best rarely equaled that of other children in the twins' respective schools – despite the home environment of one twin in which his parents worked hard to overcome his limitations.

Yet not only do identical twins perceive, think, and move through the world in similar patterns, they do so with a similar underlying pattern for growth – so that the timed emergence of these maturational faculties and their function is also shared. Although the growth progression of maturational features is not necessarily the same, meaning that one twin might have a spurt in speech ability, for instance, while the other lags behind, by and large identical twins do tend to converge in ability at some point in growth or maturity even if they deviated earlier – suggesting a ground plan for maturational features.

In cognitive ability, for instance, we found particularly fluid times, especially around four or five years of age and just before puberty, when one twin leaped ahead of the other, the network of his brain coalescing with outside conditions in some as yet unknown way to produce a leap in ability. His twin might be in an understimulating environment, needing but not getting special attention in order to move ahead – and yet both twins ultimately converged in cognitive ability. This result is reflected in an old study that today seems crude, though it remains a good illustration. Theodosius Dobzhansky describes it, quoting from the original report by Gesell and Thompson.

When identical twin girls were between forty-six and fifty-six weeks old, one of them was "subjected to hundreds of hours of preferred and specialized treatment training designed to improve her motor coordination, her neatness, her constructiveness, her span of attention, her vocabulary." This training, while it lasted, placed the twin so trained appreciably ahead of her "control" co-twin. When, however, both twins again received the same treatment, the control co-twin progressed at a faster rate, until she equaled the other.[2]

The fact that the rates of maturation may be uneven at times, proceeding in leaps rather than gradually, is a valuable notion to keep in mind. At an arbitrarily chosen moment, one girl's language skills may run ahead of or lag behind her motor ability – and ahead of or behind her identical twin's language or motor ability. Eventually a leveling-out does occur between twins and nature is revealed, but until that time we must be aware of how much individuals differ according to the blueprint of their own internal clocks, not just their environments. These clocks predispose them to change and to periods of special adaptability to the environment, called critical periods, as we will see in a later chapter.

Returning to disease, twins have shared problems such as eye disorders, orthopedic complications, and epileptic seizures, often within the same periods of time. The prevalence of infectious diseases in children – ear, nose, and throat, gastrointestinal, urinary tract, and respiratory problems – were frequently shared. Illnesses one otherwise thinks of as caused by the environment may often have genetic bases. And most interesting, the timing of these illnesses, with some variation, also seems to be shared.

> *Two otherwise healthy twins, living apart, began having spontaneous convulsions. The pediatrician of one of these twins responded by treating the source as an infection or some allergic reaction to food, whereas the pediatrician of the other twin considered a brain disorder and examined the child for lesions. Neither approach alleviated the problem. These episodes lasted for a few months and, seemingly by themselves, disappeared.*

If each pediatrician had known that his patient's convulsions were shared by an identical twin, they each might have approached the condition differently, with a strong suspicion of genetic influence. As it was, as so often happens, an inherited timetable for both illness and growth was not part of the repertoire of diagnoses.

Maturation and Disease in Adulthood

The genetic messages spurring a child toward maturity do not suddenly stop at puberty. Our maturational clocks operate within us all the time, reliably predisposing us to certain changes at each stage in the life cycle. At puberty, for example, the hormonal adjustments within the body are timed: identical twin girls separated at birth often experience menarche (begin to menstruate for the first time) within weeks of each other. A disruption of the menstrual cycle may also be inherited:

> *Olga and Ingrid were identical twins separated at two months of age who lived their formative years without any knowledge of each other's existence. Ingrid, in fact, didn't learn about her twin until they met for the first time at age thirty-five. They then discovered that when they were both between eighteen and nineteen years old they had stopped menstruating. Since they both had boyfriends and were sexually active at the time, each assumed she was pregnant and got married. A few months after the weddings, however, their periods resumed.*

This disruption of the reproductive cycle without pregnancy might reasonably be viewed as the sign of either a physical or an emotional problem – which one depending on further tests.

When identical twins reared apart both reveal this problem, however, biology becomes the more likely cause, especially since the timing of their menstrual disruption was so similar.

In another example we can see that testosterone levels influence the surging aggression and rebellion in adolescence, and their leveling-out in the twenties (along with that of the pituitary hormone) aids in the relative peace that leads to the potential for mastery of a profession, as outlined in Erik Erikson's epigenetic system. The processes of aging are also initiated by genes: hair loss, myopia, a gradual diminution of sexual response, senility – these and other phenomena of maturation are biological benchmarks, and their timed emergence depends on appropriate genetic messages.

We are by no means the first to consider that a dispositional readiness for disease or disruption is also linked to the immune system and the genes governing that system. An illness caused by bacteria or a virus may also be seen to be caused by an inherited immune system and a susceptibility to particular problems. This process may be analogous to the way we get certain cancers, breast cancer for one, which is stimulated by the presence of a gene called an oncogene. We may be susceptible to cancer if the normal genes that suppress the oncogene are themselves suppressed and the oncogene becomes activated. There is some evidence that the cancer-preventing genes, called "anti-oncogenes," are located on chromosome 17. Exactly how that chromosome becomes damaged is not known, but it may turn out that other cancers, such as colon and lung cancer, are also prone to emerge because of it. Genes may therefore create a *tendency* to pathology rather than any one disorder itself.

Alzheimer's disease is encoded in the genes at birth but only becomes expressed in late adulthood. Its timing and course are sadly predictable. Heart disease too may have strong links to a timed heredity, with some people predisposed to earlier trouble than others. The role of inheritance in psychoses is also becoming clearer. A team led by Janice Egeland has revealed (tentative though they still are) fascinating concordances of manic-depressive disorder in an Amish population, and numerous adoption and twin studies – as well as chromosomal mapping – have shown that schizophrenia is strongly linked to inherited patterns and the chemistry of the brain. For example, schizophrenics are far more likely to have schizophrenic family members than are people in the general population, even when their relatives are raised by healthy adoptive parents – and a schizophrenic's identical twin will have an enormous statistical chance of encountering the disease himself, often coinciding in timing with that of his sibling.

Logically enough, if four thousand diseases are already known to be linked to our genes, it stands to reason that longevity itself tends to run in families. We all die from something, even if it is simply "old age," in which case our avoidance of most of those "somethings" may be predisposed. Any doctor would advise that for the best chance at long life one should pick a long-lived family: it beats even the most modern medical techniques.

The Influence of Environment

Of course, the surroundings in which maturational features emerge can alter both form and function – and, for that matter, determine whether they appear at all or even whether the organism survives long enough to experience them. The genetic timetable we speak of is certainly more a rough guideline than a train schedule. It is a plan with a specific range of flexibility built into it, flexibility that awaits signals from outside. In the last century, for instance, improved nutrition has lowered the average age of menarche by several years, from an average age of thirteen or fourteen down to ten or eleven. Good nutrition and health seem to spur on the body's reproductive cycle. If one girl is fed a much better diet than her twin sister, she might start menstruating sooner, but these twin girls, sharing the same genes and the same ranges for expression of these traits, will no doubt be closer in the timing of their menarche than two unrelated girls. And continued improvement in nutrition will not lower the age of menarche indefinitely. A supremely well-nourished girl will rarely begin to menstruate in her seventh year. There are built-in limits to all flexibility in the genetic clock.

Life experience also plays a real part in the expression of hereditary illness. Even if the range of variation for a disease is quite narrow, implying that very little or nothing can be done to stave it off, fluctuations are possible. In Alzheimer's disease, for instance, although a genetic link is a virtual given, some studies reveal that 20 percent of all sufferers may have had some kind of brain injury in their past, suggesting either that the disease may be induced in other ways or that brain injury may hasten, delay, or otherwise change its presentation. Alzheimer's patients also tend to be more depressed than normal groups of the same age. Is this characteristic a result of the disease, or does the brain chemistry of depression facilitate Alzheimer's symptoms?

The incidence of heart disease is another example of how environment modifies the expression of genetically timed traits, for the risk of it, as everyone knows, can be reduced by appropriate diet and exercise. Still, if the causes of heart attacks were seen only from the environmental side – if they were seen only as the result of too much cholesterol – then those at greater risk because of a family history of hereditary cardiovascular disease would never be identified and given proper advice on how to avoid it. If genetic timing were completely ignored, we would all follow the same general advice on prevention, failing to recognize that some people must be even more careful and follow stricter diets than others.

Such a case provides great reason to understand the interaction of genetic timing and environment, an interaction that changes over the course of a lifetime. It changes not only because our experience and environment change but also because genes switch on or off at particular moments. By taking into account an inherited inclination toward disease in addition to a specific inherited disease itself, one can begin to consider the range of susceptibility to which we are individually predisposed.

In general, coming to know maturation in the whole cycle of life helps us recognize the features that maintain themselves over time and those that are more rigid or more pliable, as well as the timing of greatest rigidity and pliability. We may then begin to see the relative weight of heredity and experience on each. Here are three final examples, which serve to illustrate not only maturational effects but also unavoidable patterns of development.

A girl, Frances, had a powerful striving for physical movement. Frances's mother, in fact, reported how active she was even in the womb. By two years of age she propelled herself into constant action, exploring her environment and employing her intense motor abilities before she was able to judge the consequences. Because she would run around wildly without being able to assess the dangers, she was called accident prone. It didn't deter her. Yet when she moved beyond a certain circle of safety, she would suddenly become anxious and cry. This happened when she moved farther away from her mother than she could tolerate, that is, farther than what was appropriate for her development. In sum, her maturation in the area of mobility at this time exceeded the rate of her emotional development.

Child analyst Margaret Mahler refers to this stage as "rapprochement," and it is characterized by the way children check, look, and listen for their mothers even as they wander away from them.[3] In this rather pronounced case, Frances moved so far away that she often found herself lost and alone, where she would feel disoriented and anxious. Her mother could have left her at these times, letting her pay the price, in a sense, for her advanced motor ability; instead, for those months before her emotional development caught up with her physical maturation, her mother tried when she could to follow Frances around and thus minimize the danger. Adjusting herself to her daughter's growth, exhausting as it was, provided the flexibility in nurturing that her child's individual rates of maturation and development needed at special times.

Ellen was a four-year-old girl whose ability to speak lagged well behind her capacity to understand language. She tried to communicate with gestures and sounds but became quite frustrated when her attempts to give cues about her needs were not correctly read by her parents. This went on for a year and a half until she slowly began forming words. Then, in a rush, her ability to communicate verbally equaled her ability to understand, and she progressed quite well.

This form of imbalance can frequently be seen in children, and it too demands special flexibility on the part of teachers and parents, a flexibility to be patient with nature's rates of growth.

There are other instances, however, where some developmental dimensions never quite catch up to those of maturation. It is interesting that although the physical apparatus of the inner ear is fully adult size ten weeks before birth (and the eyeball nearly four-fifths adult size), the infant is almost emotionally helpless at that time. Physical maturation may precede emotional development, or conversely, a child may understand words and the emotions that go with them long before he is able to communicate his response in physical speech.

Surely there are norms for maturation – where a child "should be" at a certain age – and when one or more areas are lagging behind, guidance or intervention should be sought. In a general sense, however, the individual matures at a pace that is genetically his own, echoing his own emerging individuality. Nurturing will have more influence during certain times in the life span, at other times less, but the sensitive parent, spouse, or child will see that slower areas of growth may in time become equal with others, or even advance beyond them; the task is to recognize the individual's rates and modes of growth and give sustenance to them.

> Lawrence's physical functions had become reduced with age, although his mental abilities were still mobile and clear for a man of eighty-seven. A number of his younger friends were somewhat more active than he, and so he often became sad and bewildered that he could no longer use his body to follow social interests, such as holding cards steady in his hands or taking a bus with his friends to the theater. But his daughter taught him compromise methods of coping, replacing some tasks with others and at least partially satisfying his purposeful agenda.

Lawrence's daughter knew him, who he was and what he had, and she knew how to help in ways specific to her knowledge of his particular pattern of aging. This type of knowledge, at once specific (that certain maturational faculties and their timetables are linked to the genes) and flexible (that each person's internal pattern will have its own unfolding within life's changing conditions), can be the goal of anyone looking to understand himself or others.

Recognizing that a person has some early, genetically endowed traits is not enough, for there are always shifts in development. Each unfolding life is the individual variation on the general themes of maturation and development, the changes that are bound to occur over time. Understanding the individual timetables of growth may thus help us define our approach to each person, assisting us in recognizing each pattern of his unfolding and helping us adjust our own timing and needs to his. It is our ability to attune to him and to harmonize with him that enables us to help when help is needed but not interfere too strongly or impose too many verdicts on healthy development. We might even say that a parent's or teacher's most basic obligation is to discover what each child needs in order for him to become who he is. We are reminded of Tennyson's *In Memoriam*:

> The baby new to earth and sky,
>   What time his tender palm is prest
>   Against the circle of the breast,
> Has never thought that "this is I":
>
> But as he grows he gathers much,
>   And learns the use of "I" and "me,"
>   And finds "I am not what I see,
> And other than the things I touch."
>
> So rounds he to a separate mind
>   From whence pure memory may begin,
>   As thro' the frame that binds him in
> His isolation grows defined.

The vision of a growing individuality applies as well to adults as to "the baby new to earth and sky." Whereas some characteristics of each "separate mind" become accentuated late in life, others diminish. Whereas some are merely habits grown firmly embedded, others represent the programs or phases of aging.

It is this consideration that forever demands our attention, continually engaging us and giving meaning to our participation in the growth processes of life. Just as we have learned much about environmental pressures on people, the more we learn about genetic influences over time the more intricate, specific, and fine-tuned our own relationships with others can become. Instead of a wide-open approach to development where anyone can do anything, an awareness of an individual's particular unfolding – both its restraints and its potential – gives every relationship new meaning.

Though this awareness does not always come easily, the search for it should be constant, for the discoveries along the way will reward any effort. And this awareness may be heightened still more as we consider a whole range of other, more psychological features, namely, those embedded in the process of development.

Chapter Five

Timetables of Change: Development

He allowed himself to be swayed by his conviction
that human beings are not born once and for all on
the day their mothers give birth to them, but that
life obliges them over and over again to give birth
to themselves.

Gabriel García Marquez,
Love in the Time of Cholera

L ife didn't treat Ebenezer Scrooge in the kindest way. The miserly, disagreeable fellow we remember was in love once, Dickens wrote, until the world of business took him over, his fiancée left him, his sister died, and age put the finishing touches of hardness on the man.

We know what happens then. His bookkeeper, Bob Cratchit, suffers horribly in his employ: fifteen shillings a week, barely enough to feed a poor family, what with Tiny Tim's health so wobbly. Scrooge needs a lesson: a Christmas Eve to shake the soul, a terror-filled confrontation with past, present, and inevitable death, and haunting spirits to refresh his memory of what it means to be alive. The lesson is well learned. The Cratchits, thanks to Mr. Scrooge, eat for Christmas that year a turkey that "never could have stood upon his legs, that bird. . . . Twice the size of Tiny Tim" – and this just for starters.

If we take Dickens at his word, the benevolence emanating from Ebenezer's heart at the end of A Christmas Carol represents the man as he will stay – and the man he always could have been, too. Christmas Eve didn't change him so much as it returned him to himself, for in some way his transformation had to be a natural return – one he was prepared to make – or else it simply would not have occurred. Harsh circumstances may have hardened him, but Scrooge, Dickens might have said, always had a small flame burning in his heart, and that flame, part flexibility, part empathy, part of who he always was, revealed itself again with the help of a nightmarish push.

Experience is believed to have the power to change us. We acquire new wisdom, seek out new facts, are prey to new follies, and consciously and unconsciously encounter new stimulation along the way, all of which changes brain and behavior patterns. Experience has both breathtakingly powerful and minutely subtle effects, helping or hindering development depending on our ability to adapt to it. Prior experiences affect present ones, and the present affects the future.

At the same time, the image of the rolling snowball of learning is a poor one. People do not go out each day acquiring more and more bits of information, processing them regularly until they have accumulated such a large sum of data that conclusions and decisions naturally appear. What we bring to the world we experience is as significant as that which we encounter. What we see is shaped by how we see, and how we see is influenced by our genes. The infant does not learn of the mother's breast only through his experience with it: he is programmed to react to it and suck by instinct. He is not the passive receiver of information; he searches it out actively in ways and to degrees that are his own, and he does so not just during the years of education but all through his life. Scrooge was not merely a passive receiver of life's lessons; he had a naturally active hint of empathy inside him (and even empathy, we have seen, is linked to the genetic matrix), and a persuasive push returned him to that older, natural, more empathetic self.

Given the will and the right conditions for insight into our predisposed patterns, nature will return to us – or we to it – even if it is not the same as it was when we were young. We thus come to a central issue in understanding nature and nurture, namely development, the process of becoming characteristically oneself. Because of its complexity, it may be difficult to see development in any way other than from the side of experience. Modern research, however, now gives us a window.

Maturation, as we saw in the last chapter, encompasses all the biological benchmarks of growth through the changing faculties of perception, cognition, and sensorimotor skills. In contrast, development consists of a blossoming psychological landscape from which character, judgment, value systems, a capacity for relationships, a sense of reality, and a sense of self emerge. These features, even more than those of maturation, help create a pattern of both behavior and personality that from birth through maturity and death fashions each of us uniquely. If the leash between genes and the maturational process is fairly short and tight, then that between genes and development is much longer and looser, allowing much greater leeway for the influence of nurturing. The eruption of teeth and the onset of speech are more controlled by our genetic predisposition than are the rules we develop to make judgments of people. The content of such judgments is actively influenced by the family and culture we experience. Yet we can begin to see that the forces by which we make those decisions – patterns of psychic defense, of reality perception, of relationships – arise from foundations that are governed by more than the environment in which we grow.

As Freud explained, we live at the crossroads of both inner and outer demands. On the one hand there are instinctual drives, wishes, and fantasies, which are primarily biological and pleasure-seeking, and on the other there are outer expectations and conscience, which are mostly learned from parents and society. In Freud's structural theory of the mind, it is the ego that must mediate between these two other arenas, establishing a sense of reality and employing particular psychic maneuvers that help balance inner and outer demands. The character of our resolution to these conflicts helps to configure the individual psychological package we present to the world. Inasmuch as psychological development is guided by the development of the ego, we are necessarily guided by influences that range from strict genetic to powerful environmental forces. In the next chapter we will explore more fully how inner and outer drives become reconciled through adaptation, and how some psychic defenses that aid in such reconciliation have biological roots. Here we will see how the critical process of development itself, often viewed as shaped largely by environmental influences, is linked to biology and genes as well. Let us look further.

The Developmental Sequence

Erik Erikson proposed what he called an epigenetic guide to the life span, a breakdown of the stages of development, with old patterns reorganizing into new ones at each stage along the way. Using Freud's theory of psychosexual stages (oral, anal, and genital) as a base, Erikson agreed that "anything that grows has a ground plan, and that out of this ground plan parts arise, each part having its special time of ascendancy."[1] The ground plan of which he spoke, of course, is fundamentally biological and genetic.

Like maturation, developmental stages are regulated by a combination of social and biological demands. Just as the boxcars of a train may be pushed from the back or pulled from the front depending on where the locomotive is, genes and environment push and pull and spark the movement of one stage to another. As one stage is resolved and its demands met, the next stage is begun, initiating a new set of demands. "The life cycle of any organism," Dobzhansky writes, "is a sequence of stages succeeding each other in a definite order. The order is there because the developmental events at a given age or stage are brought about by the preceding events."[2] Although the development may seem continuous, it is broken. Writes Piaget, "Stages . . . can only consist of successive steps or levels of equilibrium, separated by a phase of transition or crisis."[3] Piaget's co-worker, Barbel Inhelder, adds: "What actually does a stage signify if not a change of qualitative order, a sort of metamorphosis?"

And yet one can ask, How rigid is this order of development? Though certain dimensions of growth are hard-wired to the brain and mature through the influence of a genetic clock, is the same necessarily true of the development of behavior and emotion? Of conflict? Of personality? We do know that movement from stage to stage in development is not random. Resolutions of the problems arising in one stage are essential for a healthy transition to the next, and one of the primary reasons for the push through development is the shift of energy from stage to stage and zone to zone: it is, in sum, predisposed by nature. And in the same breath it is also parental, social, environmental. "Personality," Erikson writes, "can be said to develop according to steps predetermined in the human organism's readiness to be driven toward, to be aware of, and to interact with a widening radius of significant individuals and institutions."[4]

Granted that both genes and environment have a role in development, but what ways and to what extent does each contribute?

Observation of identical twins separated at birth has provided us with an open view of the progress of the phases of development as they unfold. By following several paths of development (the psychosexual stages of development, the separation-individuation sequence spelled out by Margaret Mahler, and the stages of ego development outlined by Anna Freud and others), we observed that identical twins reared apart share *the same timing* of their developmental phases much more strikingly than non-twins. If one twin reached a certain level of organization at a certain time, the other did not lag behind for long or progress much ahead even when the environmental pull was particularly strong for a time. This finding has been noted by other researchers as well.

In one case of identical twins reared apart the home life of one twin had a more molding influence during the

*oedipal conflict (age four to six) than did that of the
other. The first boy had a father who was particularly
strong in asserting his role, paying much attention to
the child and offering himself as a role model, not only
in how to be a caring father but also a giving husband.
The father of the other twin was frequently absent during these years (though not later), which left the boy
uncertain of his abilities and of when to act out and
when to be quiet. But this difference in itself, significant
as it is, did not permanently separate the developmental
paths of these identical twins. In fact, after a few years
both boys revealed similar progress and confidence, as
if the difficulties of the second boy had been imposed
on him from the outside but had not (yet) become part
of his permanent mental organization and development,
which continued to move ahead just like that of his twin
brother.*

Such a result may seem strange. After all, psychodynamic theory and developmental research tell us that the way we play
out the early periods in our lives influences all our later relationships, all our later days. The second boy's oedipal constellation was altered. Are we now minimizing that?

Life experience, whether stimulating or depriving,
shapes each of us individually in nearly everything we do. We
cannot avoid it, even if we turn away from life, which is itself
an experience. The notion that early experience creates patterns
we later follow is an essential one, and to ignore it counter-
mands the lessons we have struggled to learn in this century
about the importance of appropriate nurturing.

What we are questioning, perhaps, is the meaning of
the word *appropriate*.

Nurturing a child with the expectation that everything
we do takes its toll and makes its mark is akin to drawing with

charcoal and fearing that every line will indelibly survive to be seen, as though any mistake will always point to our ability or inability as parents, teachers, or health professionals to nurture a child. The error in this assumption lies in believing that people are formed only by parents and others. They are not; they are also driven by their own individual makeups. Nature leaves its thumbprint in a way that inclines us to embrace, tolerate, or rebuff the situations in which we find ourselves. We are not the sum of experiences alone. The psychosexual stages are both biologically *and* environmentally influenced. In addition to a range of flexibility there is an internal plan for development, an innate potential that pushes it onward. Caring for children appropriately must mean nurturing them in ways appropriate to each child's natural inclinations. It must mean giving sustenance not to a blank canvas that records every touch but to the child's own unfolding patterns of growth. This notion must never be thought to negate the critical importance of early childhood. Rather, it enhances our understanding of how best to sustain the natural progress that is part of development.

A six-year-old boy began to worry a new teacher at school because he never seemed curious in the reading class she taught. When she asked older teachers about him she discovered that his parents had just recently separated, and she attributed his behavior to that stressful situation. So she put less pressure on him in class, never calling on him or forcing him to take part in difficult exercises. It took another few months before she learned that the boy had developed slowly at school even before his parents' separation, and usually responded better to one-on-one tutoring. Though he never became a fast learner, by working with him once a week after school she encouraged a greater degree of partic-

ipation in class, and she accepted that, with benefits in all his other subjects.

Recognizing those areas that are more locked into inherited patterns and those that are more flexible should be part of the repertoire of child and self-study. Parents, teachers, doctors, and therapists can benefit from knowing where they can have the most influence and where, in the case of genetic timetables, they must work with nature's plan in mind. Yet in knowing the individual child, can we wait long enough for him to benefit from our special guidance, or will we try to impose it? Conversely, can we be ready to lend a helping hand when he needs it even if that need appears earlier than other children's, or will we always wait until the "average" time? And can we intervene in one realm of a child's unfolding individuality while holding back in another? These are questions we can answer only when we begin to make sense of the data that are now available about nature's ground plans for growth.

Although the precise nature of the struggle with parents during the oedipal period is co-determined by family dynamics, it is interesting to observe that the potential for that struggle is shared by identical twins. Pre-oedipal, latency, and adolescent phases are also guided by a biological plan that influences changes in both mental and physical organization. This may seem difficult to imagine. But consider a young girl who was dislocated from home because her parents were forced to flee to another country. What she most fiercely resented about the move was not the loss of a house or a homeland but rather the disruption of her relationship to her peers. Such a reaction is specific to her level of development: it is part of a sequence in relationships that reveals itself at that age but isn't visible at all in a three- or four-year-old child. Or consider the struggle for autonomy and control that occurs during the so-called terrible twos. That struggle is also age-specific. It isn't often found in a five-year-old, who is working under different conditions with a different set of issues.

In fact, there are also timetables of deviation from the developmental path – in other words, predisposed influences for neurotic conflict. We have discussed in the chapter on maturation the specific inherited diseases with one-to-one genetic links. The question here is not whether but rather how much heredity influences developmental disorders. And if we fall off course, can we return? "Suppose that some developmental events are hastened or delayed by environmental or genetic causes; will the organism be able to compensate for this disturbance and revert to the developmental path it would have traversed if the disturbance had not occurred? Or will the whole subsequent pattern be altered?"[5]

Sigmund Freud saw that each phase of development and maturation has its own set of corresponding conflicts, which when resolved lead to a new phase and new conflicts. Health is not the absence of conflict, like an absence of brain lesions. Freud saw health as the condition *that allows for* the resolution of conflicts in childhood and, later, the resolution of each person's inner strivings with his environment. Health is the ability to love and to work in spite of these conflicts.

In this sense, the fears of the five-year-old child, his nightmares, his guilt over his primitive wishes, his oedipal struggle – these symptoms are an expected part of healthy development. They lead to what Freud called infantile neurosis, which for an adult could signify abnormality but for the child of that age is a normal, expected part of growth. In this sense, one can even say that Freud "normalized" neurosis, at least to the extent that it appears and recedes as a healthy part of development – with consequences that retain power throughout life. The child's ritualistic behavior at age two or three – his endless repetitions of the same game, like peek-a-boo, or his demand for an exact performance of bedtime habits, or his wish to be told the same story word for word – these seemingly compulsive behaviors can serve the need for continuity and regularity that is completely healthy for children at this age but not later.

Of course, genuine abnormalities in development do exist; but these too can be seen to be embedded in the process of development and not isolated disease entities. As Serge Lebovici and René Diatkine write, "To speak of normality in connection with children who do not draw attention to themselves [by revealing specific illnesses] is to run the risk of overlooking hidden pathology . . . we have to raise questions regarding the notions of normality and abnormality since it cannot be taken for granted that a precise definition of these concepts will be put to good operational use in medical practice."[6]

Such a dynamic view leads to an understanding of how conflict and abnormality are part of the developmental process. Reflecting this trend, one psychiatrist has written that "'disease' is out; 'disorder' is in."[7] For what is a disorder if not a "dis-order" – a disturbance in the orderly sequence of maturation and development?

Considering abnormalities in isolation misleads us, for they are embedded in the context of a growth process that is just that, a process, guided by genes from within and environment from without. There are normal and abnormal phases and discontinuities in development, where old hierarchies and ways of functioning are superseded by new ones as growth proceeds. Although we may assume an orderly sequence, the individual seems to prefer to go his own way – producing countless healthy and unhealthy variations. However predictable phases are, each child will always have his own timing. To try to chart precise periods – weeks, months – when these faculties emerge is to forget about interaction and variability and to return the nature/nurture question to enemy camps on opposite sides of a field. Rather, intricate branching and periodic fluctuations are part of the scope of development, and we can count on both genes and environment playing their parts. One child has phases that overlap; he takes his time to develop with no clear demarcations and disruptions. Another seems to be fixed on one particular mode of functioning and can't progress to the next phase. Still another reveals the coexistence of a few phases, leaving others behind. And some children proceed rapidly through development while others go as if in slow motion.

Consider the case of this woman:

An intelligent twenty-seven-year-old who works in a law firm, Marlene described her limitless need for social contact. When she was alone in the evening after work, she phoned her friends for hours, desperately needing to hear and be heard and paying much attention to the intensity with which her friends inquired about her daily life and feelings. At her job she was similarly sensitive to her status among her peers. Though men were attracted to her, some of whom she saw over long periods of time, one alone was never enough. When she had money she went on shopping sprees, and when she ran out she asked her father for more. She never quite knew how much she spent each year. Although she liked fancy clothes when she went out to be seen, she described her room as in a constant state of emergency, needing desperate attention, because someone might come to visit her unexpectedly.

In essence, who she was depended on how she was treated. She could be a queen or a washerwoman, mentally gifted or illiterate – it all depended on how she felt others saw her. She never consolidated her sense of self; her separateness and individuation never took root. She blamed her family for her endless lack of fulfillment, her mother because she didn't know how to help and her father because he was too busy. They may not have done enough to help her, enough to overcome her own inner disorganization, and she longed for others to help her do what she herself could not.

In this woman's developmental history we can see a coexistence of phases. Her primitive, oral demands and her preoccupation with the confirmation of her femininity suggests that she never passed from one phase to another with any degree of resolution. There is no hierarchy, no demarcation or discontinuity of phases. Each phase coexisted with the others even as she passed from adolescence to adulthood. This pattern differs from those that are more familiar, such as the person who fixates primarily on oral demands and dependency, or the one who is bound to anal issues of bodily control and compulsiveness; yet any of these outcomes is possible when the order of development is disrupted, resulting in dis-order.

It was a significant step when Freud linked the neuroses, previously viewed as encapsulated entities, to the very process of phase development. In doing so he saw disorders in the context of the life cycle, of arresting or changing conditions of life, of the inherited timetables of growth, and of the critical periods of special flexibility. Since the developmental sequence provides the bedrock for this process, the phobias, compulsions, obsessions, and anxiety states can be said to arise from a dis-order of development, linked to our genetic ground plans and predispositions yet sparked and malleable by offerings from the environment.

Since we have already seen how identical twins reared apart generally share the same developmental patterns, it may not be surprising to note, then, that they also often share the same disruption of these patterns.

Peter and Palle were twin Dutch children reared in separate homes and unaware of each other's existence until they reached the age of twenty-two, when they met for the first time. Described as "pedantic and limited in insight and emotional responsiveness," both boys appeared to be quite nervous as children. "Both suffered from nightmares and picked at their nails. Palle also picked his nose and had nocturnal enuresis [bed-wetting] and stammering. Peter plucked at his eyebrows when worried about exams in later life. Both were restless, nervous, and troubled in adolescence, and each chose an older man as mentor, a fact that led Peter's mother to accuse him of homosexual tendencies.

"Palle described vague sensations that had been with him since puberty. They consisted of a feeling of internal restlessness lasting for a few hours or days. He felt 'partly dead' and had sensations of soreness around his larynx. He felt his voice was 'wrong' at these times, and he mumbled and spoke indistinctly, unable to correct himself.

"Peter also described vague sensations that had been with him since puberty. He felt 'dead,' a feeling that would last for a few days before going away. Sometimes he felt he had a lump in the throat localized around his larynx. He felt his voice was 'pitched wrongly' at these times and was unable to stop mumbling and speaking indistinctly."[8]

Identical twins raised apart share many of the same emotional disorders, such as fears, anxieties, abnormal dependencies, thumb sucking, sleep problems, and others – just as we know they share psychotic illnesses such as schizophrenia and depression. What is particularly interesting is that the content of phobias varies from twin to twin. The witches or good fairies who visit a girl in her sleep resemble only those people she knows, especially her parents but perhaps other caregivers, siblings, and friends. The faces she smiles at during the day and recalls at night are particular to her experience alone, grounded only in her environment. The content of her particular relationships, dreams, phobias, or developmental phases is unique to her. There is no more similarity in the content of these dimensions between this girl and her identical twin than there is between any other pair of children. And the reasons given to explain these fears are also individual: one person blames a strong mother for her nighttime phobias, the other blames a powerful or absent father for her fear of animals.

But the prevalence of the fears is shared, whatever their content. In the University of Minnesota study conducted by Bouchard and co-workers, three sets of identical twins reared apart shared some kind of fear or combination of fears – of water, height, or containment. Environmental influences can modify, stall, or even eliminate these fears and anxieties, yet the tendency to have them, the endowed readiness for an event to trigger a new fear or reawaken an old one, is present in certain people and not others.

Peter and Palle appear to have been vulnerable in just this way. Nothing put them at rest, and the fault lay not entirely with their environments, as might be claimed for nontwins. Both men were clearly troubled in similar ways, the tendencies toward which were more dramatic than in nontwins and endowed from birth in their patterns of development, functioning throughout their lives.

Of course, problems like anxiety over separation or phobias are most always expressed within the folds of relationships. Since they might not be seen at all outside the context of parental, sibling, or peer negotiation, we can ask whether problems in development are caused by these relationships, as is often supposed, or whether they are merely expressed in and by them.

With an awareness that a genetic timetable for development and conflict does exist, the relationships that were once thought to be the sole cause of these conflicts may now be reappraised. We can see that the interaction of parent and child is not the only cause of every emotional difficulty. Perhaps it is only the most recent cause, for the child brings to that first relationship an inclination or susceptibility to order or disorder *within him* that is not the sole result of what his mother or father does.

Our lives turn on events we remember and forces seemingly external to ourselves with which we have to deal – finding a job, a mate, appropriate lodging, a good doctor, schools for our children, safe water, and good enough food. We are not often aware of the inner developmental patterns we bring to these tasks, for they often seem unrelated to them, inconsequential. Yet it is our opportunity, not only as parents but also as member of a species of mammal with the incredible capacity to ponder our own existence, to make sense of the development of our lives as they unfold, each of us individually. Our ability to handle some of the hardships of life and to make use of the special love and attention that become available depends on who we are – on our internal plans for growing and being and on our varied experiences. We thus come to a most elusive and interesting issue, the multifaceted bridge that joins our individual "givens" to our individual environments – a bridge we can call adaptation.

PART II

Bridges to the World

CHAPTER SIX

The Bridge Called Adaptation

. . . This strange eventful history . . .

SHAKESPEARE, *AS YOU LIKE IT*

A picture of the relationship between our genetic plans for growth and the environment that nurtures them is much more intricate than that of a strawberry being sunned and watered. People do not passively present their needs and their dispositions at birth to be nurtured ever after. Rather, there are developing interconnections between us and our environments that characterize each day of our lives. As we look closely at this relationship, its nuances reveal the very process of our accommodation to the world we grow in; it reveals how our makeup unfolds and adapts through the process of living. This chapter is about that elaborate interconnection, that bridge between the self and the world.

Adaptation, in general, comprises all the psychological adjustments we make to the environment for the purpose of maintaining our ability to function in it. Although the study of adaptation reaches as far back as Aristotle, for our purposes we might put a simple date of 1939 and the publication of a book called *Ego Psychology and the Problem of Adaptation* written by the psychoanalyst Heinz Hartmann. Hartmann says, "Human action adapts the environment to human functions, and then the human being adapts (secondarily) to the environment which he has helped to create."[1] It is worthwhile to keep this thought in mind as we proceed.

We know from the past two chapters that both maturation and development do not follow straight lines from birth to death. They describe changes. One might even imagine growth as a line notched frequently with biological and environmental marks, like railroad tracks crisscrossed by ties and marked by occasional train stations.

Eventful indeed is each individual's history, and the changes that appear throughout life come in assorted forms. Some are marked by milestones of both biological and psychological change, as in puberty or aging, or even in an infant's sudden reaction to strangers at seven months of age. Others constitute whole groups of functions, such as the series of psychosexual phases described by Freud or those of cognitive development outlined by Piaget. Still others are slow micro-adjustments to other major changes, such as a child's honing of vocal tone as he comes to formulate words.

As we have also seen, many marks of maturation and development are "hard-wired," predictable because of the set sequence of genes that initiates very real biochemical changes in the body. A fair image of growth must account for these changes, and molecular biologists may in the not-so-distant future uncover their precise genetic coding. Except with the most hard-wired features of growth, however, a level of flexibility always exists. It must exist, for the sake of the individual and of the species. There is a degree of interplay with the environment that not only results from the process of living but is also expected and included in nature's plan. Although the bicycle that knocks us flat in the street is an example of the environment taking momentary control, we are rarely so passive. Flexibility itself is built in, and through it we can see how our adaptation to the world is genetically grounded.

Imagine first what would happen without built-in flexibility: suppose a child were genetically programmed to suck only on his own mother's breast and on no one else's – programmed to respond only to the warm scent of his natural mother and not even to a specially scented warm plastic nipple. If his mother became too ill to breast-feed, or worse, if she died at childbirth, the child would then refuse all wet nurses and bottles and ultimately all nourishment. Unless force-fed, he would die.

Fortunately, though, there is a wide range of acceptable breasts to the newborn. Evolution through natural selection did not produce infants who must drink from the mother alone. If such narrow-ranged babies ever existed in our evolutionary history they would not have survived as prosperously as the more flexible, more adaptable breast or nipple sucklers we have become. We survived because most breasts are quite acceptable – most bottles enjoyed and emptied. Although we may prefer mother's own, it is to our "natural" advantage to adapt to a range of opportunities, and *Homo sapiens* are superb adapters. Compared to many other animals, in fact, we are quite flexible creatures. Some hummingbirds respond innately to special red rather than white flowers because they have evolved beaks physically suited to those red flowers, beaks so sharply curved that only the curved openings of those particular flowers allow them to feed on enough nectar to live. If those red flowers die, so do the curved-beak hummingbirds, who have become too specialized to change over to some other bud. For the most part, people are not hummingbirds. Though we have an inherited ground plan for growth, the life story we play out is not scripted sentence for sentence. Human beings as a species have developed a possible range of flexibility that allows for adaptation to the world, and every one at birth has his own individual version of that general range.

Some children and adults are more adaptable at birth than others, some more adaptable in certain areas than others, and some at certain times in their lives and not others, as we will see. And some may even be too adaptable, with serious consequences in the formation of their identity. These ranges in our individual capacities for adaptation are as much a personal characteristic as any other; and they in turn affect all others. Our individual ranges of flexibility are like individual lens shadings, different for each of us and changing as we grow. They change not only because the environment creates new conditions for living but also because they, like the sequences of growth they follow, are linked to inherited ground plans.

Before looking at the particulars of this process, the actual case histories in which both wide and narrow individual ranges of flexibility are expressed, we might turn briefly to some of the theories of adaptation.

Communication with the environment exists at many levels: at the level of the molecule, the cell, the species. Our focus here is on the level of the organism, the individual in the years of childhood.

Jean Piaget painted a most vivid picture of the child's ever-changing world, describing a continual transaction with the environment throughout early life. We mention his theory not for its details so much as its intimate regard for the process of exchange.

His specific concern was with cognitive organization – how we come to know what we know – and so he detailed the distinct stages of development, from sensorimotor (birth to two years) to formal operations (age twelve to fifteen), when at last we are able to engage in abstract, hypothetical reasoning. As a child passes from stage to stage, he constructs in his mind complicated representations of the world outside. Although these mental constructions are subject to change, the child doesn't wait for change: he is malleable but not passive. "In order to know objects," writes Piaget, "the subject is always required to act on them, i.e. to modify them in one way or another."[2] According to Piaget, then, the child interacts cognitively with the world in two particular ways. By *assimilation* he incorporates an understanding of objects from the environment into his mental world, and by *accommodation* he modifies his mental world to fit those objects. In this ongoing relationship, the two processes are barely separate. There is a constant exchange between the child experiencing an object – taking it into his mental grasp – and modifying his understanding of it to fit the outside world. This lively give and take provides him with an understanding of reality that is both a subjective view of things and also an objective sense of what really is.

Assimilation and accommodation are like a perpetually active crosswalk between the child and the environment. If all goes well, by his early teens each person will have reached a state of equilibrium, which is the purest expression of adaptation to the environment.

Anna Freud, elaborating on her father's work and forging her own, described a similar interaction of inner and outer worlds but on an intrapsychic level. We have mentioned that the ego's mechanisms of defense mediate between inner drives, wishes, and fears on the one hand, and the reality of outside demands and conscience on the other. There are dozens of such defense mechanisms, from repression to sublimation, and we all employ them in patterns that are distinctly our own, useful to us in everything we do. In fact, an analysis of how we set up our own patterns of defense plays an important part in the psychotherapeutic process, for though the defense patterns we crystallize in development may be benign, they can also become like blocks of stone, impeding our enjoyment of life until we gain insight into them. For Anna Freud, who was especially interested in the ego's development, two of these defense mechanisms, *introjection* and *projection,* contribute to a process that permanently links the child to his human environment.

Introjection allows good objects (people or parts of people) to be incorporated into the psyche, whereas unpleasant features are expelled back onto objects in the world through the process of projection. Like Piaget's assimilation and accommodation, these two processes are continuous, sometimes even simultaneous. She writes, "When a husband displaces onto his wife his own impulses to be unfaithful and then reproaches her passionately with unfaithfulness, he is really introjecting her reproaches and projecting part of his own id."[3] In another case, the child introjects his mother's "good" qualities – those that satisfy his needs – while expelling onto others the "bad" features he cannot tolerate. This protects his image of his mother and keeps it always gratifying. Often enough these two defenses are a healthy and necessary part of ego function. Most important, they, like Piaget's concepts, reflect the perpetual communication between individual and environment.

Like Anna Freud, Heinz Hartmann was a psychoanalyst who expanded the classic study of unconscious processes beyond what is defended against, namely the drives, to the ego's defending actions themselves. He was thus interested in the adaptation of biological forces to the environment, and he believed that the ability to adapt is itself biologically programmed. In Hartmann's view, adaptive maneuvers prevent neurosis, and even the avoidance of certain experiences can be adaptive if it promotes healthy functioning. A child who shrinks from confrontation in school may be acting in an adaptive, healthful way by keeping his vulnerabilities out of dangerous situations (dangerous, at least, for him). Of course, adaptation may lead to other, better possibilities for action than avoidance. For Hartmann, adaptation is the "possibility *to search for* a favorable environment and for effective action. . . . We call a man well adapted if his productivity, his ability to enjoy life, and his mental equilibrium are undisturbed" (italics in original).[4] The environment we select is acted on individually, for each child responds differently to each parent, and in the process remakes each one to suit his needs.

It is to the child's supreme advantage, then, to have a wide range of adults from whom to choose. One implication of this notion concerns the single-parent home, which is growing more common all the time but which may not provide the variety of opportunities a child needs. With fewer people in his path, his ever-changing needs during maturation and development may go unnoticed or unanswered. Such a child may be helped a great deal by a parent sensitive and responsive enough to him to provide a wide circle of family, friends, and schoolmates in his life. We should also remember that parents too search for an environment that best suits them, often seeing the child they wish to see – and slightly remaking him in that image. What is clear in this picture is how much the process of adaptation implies activity on the part of every individual both to adapt to the environment and to change that environment to suit his needs. We can now turn to the real-life situations in which this fascinating interaction is played out.

As mentioned, most built-in ranges of flexibility have limits that are shared by other members of our species: biologically, for instance, we cannot survive when the air temperature around us reaches 200°F or falls to −200°. The range for comfort, however, is not the same in every person, nor is it necessarily the same in one person throughout his life. Ranges change, flexibility increases or decreases, and a person's capacity to adapt to his environment may prove more or less successful over time as he changes or environmental conditions change. These variations exist just as other human variations exist; they are part of the equipment of our species, and the version each of us possesses is similar to but unique among those of all human beings. Here is an example of how our adaptive capacities differ at different ages.

> *Under the threat of violence, a well-to-do family in a Latin American country was forced to flee to the United States. The parents of this family found their new life – without the overarching threat of death, disappearance, and torture – profoundly better and easier than ever before. They were relieved. At the same time each family member had to become accustomed to new day-to-day life styles, and not all of them managed with the same degree of ease. The eldest son encountered difficulty in his new school, a middle daughter missed her friends inconsolably, while the youngest boy seemingly made the transition away from friends and nursemaids*

without complaint – except that he was upset about his new bedroom, which in its arrangement of furniture was nothing like his old one.

Relocation from one city or country to another may be laden with stress, even more so when it happens under violent threat. But why does one person manage so differently from another? One might at first wonder whether our perceptions of such a relocation are different for each of us, and if our readiness for it partly accounts for the ease with which we adapt. One could expect that preparedness is something we learn as we grow older, accounting for the fact that experienced people handle problems more easily than novices. This much is almost obvious.

Even more important, however, is the fact that people's ranges of response shift with the various demands of the life cycle. The dislocated family in the example above consists of children of varying ages, and each age has its own set of conditions and conflicts. What for the middle child was an impossible situation for her age – namely, losing her friends – was for the youngest boy mostly a secondary worry, one unrelated to his developmental level. The oldest son was unmoved by the change of beds – contrasting with his young brother's anxiety over the disruption of ritual – but he was disturbed by the demands of adulthood, which included making his way in school and getting a job afterward. As internal demands and external conditions change over the course of a life, they must be met by a psychological landscape flexible enough to cope with them.

We as a species differ from most other animals in the length of our physical dependency after birth. We are virtually helpless for years, and as a result we are well-integrated into the culture and family in which we are born. We are thus well served by dependency. Since we have a long time to be parented, we learn. We learn to mold to the specific conditions we meet, our particular parents and culture, the particular foods offered us, and so on, and we emerge from this period sufficiently acclimated to our new world to survive in it. The idea of *critical periods* in animal learning applies in many cases to human beings as well. Just as there are early moments in the life of a greylag gosling when it may accept a person as its mother, following him or her around with apparent devotion (whereas an adult goose afraid of people might fly away) so too are there critical periods in human lives, times when it is advantageous for us to be flexible.

These times do not only occur in infancy. Each stage of growth is defined by its period of special adaptability, times when the benefits bestowed by the environment may be best appreciated and incorporated. A child who we try to teach to read at three months of age may get something from the ex-perience – not much, but something, perhaps only the happy sensation of interacting with caring adults. But when the child is between three and five years old the same reading lesson may create visibly exciting results, with the child forming sounds by looking at letters on a page – and from the sounds words, and from the words sentences, and suddenly the child is able to read whole books by himself, none of which would have been possible a year earlier in maturation. And although the potential for reading stays with us throughout life, the fifty-year-old adult who never learned to read will have a much greater struggle learning for the first time than the four-year-old child. The same is true in learning a foreign language, playing a mu-sical instrument, or making a major discovery in physics. (Most breakthrough discoveries, Galileo notwithstanding, are made by scientists in their twenties or thirties, not their fifties.) In contrast, the writing of poetry or prose may benefit from the associations of experience made over a lifetime.

From infancy to maturity to old age, each phase of

growth, to quote Erikson again, "has its special time of ascendancy," and we can say the same for adaptability. Certain times are characterized by the widest possible capacity for adaptation, while before or after these periods we may be relatively rigid. A predispositional readiness to be flexible is a foundation on which later development rests, and one can easily see how some people make the world meet their needs and answer their calls, whereas other people, no matter how hard they try, never seem comfortably situated and in harmony with their environment. Psychological health is the ability to make these changes.

> *For the first time since they were separated in infancy, identical twin men, twenty-two years old, met quite by chance. They sat and talked about their past and present lives for a few hours, both stimulated by the sensation of meeting a missing brother, especially someone who looked so similar – even in the cut and style of hair. After they had parted, one of the twins, when asked when the two of them would next get together, just shook his head. "He smokes and drinks a lot," he explained. "So did I, dangerously, until I got counseling and managed to quit for good. But I still fight the urge sometimes. My brother says he'll never quit. So I won't see him anymore. The urge is in both of us, you see, but he won't stop. How can I stand to watch that?"*

What is interesting in this example is not only the shared addiction for tobacco and alcohol, which one of the twins managed to control and the other couldn't, but also the wish on the part of the recovering twin to stop seeing his addictive mate. Since quitting he had surely come in contact with people who smoke and drink, dealing with them well or less well. But seeing someone so like himself – a brother, a twin – with the same urge to light a cigarette and take a drink (but without the ability to control that urge) disturbed him too much to go on meeting his brother. His method of coping with the situation, then, was avoidance, which Hartmann indicated is not necessarily ego restricting and negative. It may be an acceptable if not ideal method of adaptation, one this twin regarded as necessary given his own previous addiction and struggle to quit.

One could say of course that his decision was too extreme. It certainly was not the optimal one, cutting off contact with a long-lost brother. At the same time there are worse scenarios. By seeing his brother, his own tentative resolve to stay cigarette- and alcohol-free might start to weaken – although he might then deal with these new urges in even better ways. Yet what if he were addicted not to liquor but to heroin? Though severe in this case, his decision might be considered wise. To paraphrase the Wizard as he advises the Cowardly Lion at the end of *The Wizard of Oz*: Some people say that those who run from danger lack courage; others say they possess wisdom.

Making a judgment about whether a behavior is maladaptive is not always easy, but as we look to a person's characteristic pattern of adapting we can then assess a given moment. We can see that some people manage well only in certain environments, and some not at all during particularly troublesome times. During periods of change, for instance, some who otherwise cope well with the world find themselves confronted by environments that simply do not fit, and their behavior barely conforms.

A fifteen-year-old adolescent has again been asked to leave her private school. It will be her seventh change. She is disrespectful to her teachers, makes few friends her own age, and in fact allies herself only with older boys or adults. She pays no attention to her studies but is deeply interested in painting. Her early history reveals

*a usually happy temperament and a precocious ability
to acquire language, an excellent memory, and an extended vocabulary. When her intelligence was tested at
age eight for entrance to the first of her private schools,
she scored in the highest possible range, but after acceptance she soon became bored and stopped doing
homework. Although she received A's on exams, her
teachers felt she didn't deserve them because she never
worked hard. She soon became openly disrespectful.
Not esteemed herself, she esteemed no one in return.
Her older male friends led her to sexual experimenta-
tion. Her parents, confused about what to do, alternated
between indulgence – "as long as she get A's" – and
punishment. Her father, an ambitious man, was proud
of her and deeply upset about her difficulties. Her
mother tried to deny the problems, making her daughter
feel she was too weak to count. Lost and maladapted,
she turned against those most close to her.*

An easygoing person in childhood, this girl apparently found
herself in situations ill suited to the particular emotional and
physical transitions she had to make in adolescence. After high
school graduation her ability to paint eventually carried her
into a world of confidence and productivity. But since she had
not been able to adapt well in school and with friends her own
age, and was given little help to adapt better, the choices she
found herself making at that time led her into a spiral of ever-
decreasing opportunity. Fortunately, by the time she reached
adulthood she still possessed enough ego strength to make use
of her talent for painting, which soon opened doors for her.

Change is inevitable. Some conflicts are so compelling
that they demand from the sufferer new, unconscious methods
of coping with them. Some phobias become transformed,
psychically reorganized. A boy may start out early in life with

a fear of water, but when he reaches the age of eleven the fear seemingly disappears and a compulsive behavior emerges, such as hand washing. Is this child's emerging compulsion necessarily a different disorder from the original phobia? Or could both problems share the same root but with divergent expressions? A disorder may appear constant in one person but become transformed in another, or one person may manage it much better than another. If a boy overcomes his dyslexia at an early age and goes on to become a professor of literature, do we say that the dyslexia was never strong in him, or that his assertiveness in contending with his reading problem helped him surmount it?

As we know, some studies have found shyness to be a trait with genetic origins and quite durable throughout life. Yet although one twin with a predisposition to shyness may appear passive all his life, the other may try to compensate for that passivity by forcing himself into a setting that requires bursts of energy. He may join, say, the fire department. He will in any case appear different from his perpetually shy twin, the one who does not compensate much, if at all. Shyness itself has been described by some as a compensation for other conflicts. One theory suggests that shy people actually tend to be in a hyperactive state of arousal, which they counteract by staying constantly subdued. If this is true, then a genetically predisposed pattern may appear in many forms and be transformed over time into many more.

These compensatory maneuvers are often on the side of health. The changes they bring are adaptive ones, employed because they work well for the user. As we have suggested earlier, the ability to compensate and the choice of defense mechanisms themselves may be genetically influenced. In her seminal book *The Ego and the Mechanisms of Defense*, Anna Freud proposed a hierarchy of defenses, from quite primitive ones employed almost reflexively early in life to more advanced mechanisms requiring complicated psychic maneuvers. Like other individual features, the defense organizations that the ego uses are influenced by constitutional factors. A person may be predisposed to certain ways of responding, using defenses that are most basic or that work best with his innate inclinations. In Sigmund Freud's model of the mind, defense mechanisms comprise a significant portion of unconscious mental activity with (quoting Anna Freud) "a single purpose – that of assisting the ego in its struggle with its instinctual life."[5] Notice in this phrasing that the ego is said to have its own instinctual life, its own innate ties to the forces of biology that codetermine its strength and expression. In fact, tendencies toward ego disorder or ego strength are almost always shared by identical twins reared apart and discernible in each of us if we begin to look for them.

At the same time, over the course of individual development new methods of coping may be added to the older, more primitive ones, especially if there is pressure from the outside for immediate change. Psychic organization greatly depends on the life experience of each individual, and certain people, interestingly enough, are born not with too little but with *too great a capacity* to respond to offerings from the environment. With no developed inner sense of identity or firm sense of self, such children flow too freely in every direction. Every new presence seems to mold them differently and diffuse their lives. We know of a pair of identical twins who from the first days after birth appeared to be easygoing and flexible but who, after years apart in separate homes, behaved quite differently from each other. Each reflected his differing environment too well – much too well – with the result that each, like Woody Allen's Zelig, a character without any inner strength or ability to stand for himself, never crystallized a sense of self or developed an enduring identity. These twins "became" anyone they were near. Too much resilience continued for too long, and the children became responders in life rather than initiators of their own lives. The environment shaped them powerfully because they never consolidated their own identities.

Yet life experience may also play a more definite role. A case similar to that of the Zelig-like twins could be imagined if, for example, a primary caregiver suddenly disappeared from the boys' lives. Missing the stability they once had, they might turn to anyone for help, moving from lap to lap restlessly, unable to be comforted. Without knowing the individual's history we cannot tell whether a Zelig is too adaptable because of his genetic predisposition or an environmental trauma. The issue of adaptation is almost always complex, and the best prescription for decoding behavior in these cases is, as always, to consider the individual over the wide course of development.

Here is a final example, then, a case of both son and mother, hinting in its complexity at the complexity of life.

Although nine-year-old Ethan was gifted in many areas, he was without friends at an age when friendship usually gains the highest priority. He didn't want it that way; he was painfully aware of how he was treated by others, but he couldn't avoid provoking their criticism that he always had to take center stage in class, show off, be the best. This combination of intellectual ability and emotional immaturity was compounded by his relatively restricted adaptability; unlike his peers, he stopped exploring new situations and couldn't find an inner calm around people. From early childhood on, in fact, he was irritable and depended on his parents both for their attention and their help in regulating in him what he himself could not control. His early irritability turned later to anger against his parents, but this did not diminish his dependency on them. He expected his mother to be endlessly available for him and always in sight, and he never understood her need for privacy when she expressed it. Instead he complained that she didn't understand him, which was correct if he meant by it that she couldn't help him. Neither could his father, who Ethan complained was too distant and, by implication, too weak to help him overcome his sensitivities.

In this way he projected his vulnerabilities to the outside world, externalizing his problems to those around him, making their problems the cause of his troubles. As long as he could perform this psychic feat, he protected himself from the anxiety of his failures. Later he found relative peace in the pursuit of his hobbies and talents. He liked collecting stamps of the world, fantasizing about places different from his own. He was also fascinated by telescopes, whether spying on the lives of others in his neighborhood or gazing into the universe. He learned to master computers much earlier than any of his peers: in them he could control without ambivalence and fear the rules of order and causality. In this particular way he was flexible, creating for himself the external conditions that suited his needs for peace and superiority. Perhaps his future will even bring contentment. He might follow his interests and become an astronomer or a collector of specimens and arrange a satisfactory life independent of others.

His mother's life story reveals a history of vulnerability and adaptation similar to her son's in several ways, significantly different in others. From early childhood she was plagued by fears of natural disasters, global wars, and physical injury. She experienced the world as both unreliable and threatening; even her simple survival, she felt, demanded her full attention. Yet in spite of her fears and sensitivities, which she could not overcome, she was a friendly person, and when her son was born she felt extremely happy. He was the apple of her eye, the central meaning of her life. But the sensitivities she transmitted to him matched her own, and because her own strength was depleted, she couldn't comfort him in the way he seemed to expect.

She did become protective of him (the way she was about herself, against the threat of global calamity, and so on), and this bit of maneuvering toward him — instead of rejecting him outright — helped her learn that his reactions to life's situations were as exaggerated as her own. As she learned to restrain her own fantasies, she made an effort to calm his as well. She encouraged his scientific curiosity and encouraged her husband in his role as father. She gave up pushing Ethan toward the social activities that made him anxious. Although she protested his relentless dependency on her, underneath it all she liked the feeling, for it confirmed her sense of belonging and reliability. Though accused by her son of bad mothering, she found in her relationship with him a confirmation of her significance in life.

The vulnerability of both mother and son kept asserting itself; however, the two of them managed to adapt to their situations and found, perhaps only for a time and mostly on the part of the mother, a value to their relationship they hadn't thought existed. Perhaps these particular adaptations were not the healthiest they could be, and family counseling or individual psychotherapy would surely have been of help; but they may have been the best opportunities available, especially considering the alternatives.

The life of our species and that of each individual in it depends on our ability to adjust to what we encounter. It is, as we have seen, an ability with a built-in range of expression and is a part of our endowment at birth. The periodic changes in this adaptability are part of what distinguishes one person from his neighbor. They determine the breadth and character of the individual bridges we create between ourselves and the world, letting us pursue each "strange eventful history" in our own way. We begin to recognize these individual ranges of flexibility in ourselves and our children as we follow their intricate unfolding over time. And perhaps we see them best of all in two striking examples of human variety: the vulnerable and the invulnerable.

CHAPTER SEVEN

The Vulnerable
and the Invulnerable

*Human nature will not change. In any
future great national trial, compared with
the men of this, we shall have as weak
and as strong, as silly and as wise, as bad
and as good.*

ABRAHAM LINCOLN

T he *New York Times* of October 13, 1987, carried this
story:

> "*A woman, a paranoid schizophrenic, ate all her meals
> in restaurants because she was convinced someone was
> poisoning her food at home. Her 12-year-old daughter
> developed the same fears and likewise ate in restaurants.
> Her 10-year-old daughter would eat at home if her
> father was there, but otherwise went along with her
> mother.*
>
> "*But the woman's 7-year-old son always ate at
> home. When a psychiatrist asked the boy why, he said
> with a shrug, 'Well, I'm not dead yet.'*
>
> "*After several years, the older daughter devel-
> oped paranoid schizophrenia like her mother. The
> younger daughter, while sharing some of her mother's
> fears, managed to go to college and adjust fairly well
> to life. But the son went on to perform brilliantly in
> college and in his adult life.*"

Compare this boy to one described by a woman we know:

> *My son was difficult growing up. As a baby he was
> crying more than my friends' babies, and whatever I
> did to try to soothe him or comfort him never seemed
> to work. He was too sensitive. Light and noises, the
> vacuum cleaner, thunder, flushing the toilet. He would
> jump. When he got a little older it seemed his feelings
> were always getting hurt. In school his teachers told me
> he was usually shy, which I knew, but he did have one
> good friend, a neighborhood boy who was also pretty
> shy. My son would try and take control of his friendship
> with this boy and tell him what he wanted, but it wasn't
> like he wanted to bully him. It just seemed like he was*

*trying to find his own way of getting some personal
control, only he just couldn't. And whatever I've tried
to give him doesn't help, and it's the same today as it
was when he was small. He has talents, you know. In
fact he's quite superior in some subjects, and he plays
the piano . . . but he's not happy.*

Why are the boys in these cases so different from one another?
Does it have to do with the support – or lack of it – they
received from the outside? Yet they themselves brought to the
world a tendency toward regulation or disharmony that was
within them: they were born with an ability or noticeable
inability to adapt to the range of pressures they met. These are
extreme cases, surely. But in the extreme we can begin to see
some of the effects of nature as it reciprocates with nurture.

Vulnerability and invulnerability (known also as susceptibility and resilience) are not traits as hair color or blood
type are traits. Rather they are characterizations of much more
general patterns of behavior, patterns through which people are
able or unable to draw from the world around them the comfort
and nourishment they need for development. The invulnerable
child (the boy in the first example above) managed to survive,
even thrive, in the midst of a family's mental illness; the second
boy achieved no inner sense of tolerance or regulation in his
life, regardless of the efforts made to help him.

These two terms are neither elegant nor precise. They
are also too broad to be diagnostic. (Few doctors would label
the cause of a problem as "vulnerability.") Still, they are not
hard to see. Parents will recognize these behaviors in the sim-
plest situations, such as the child who scrapes his knees in a
playground and is overcome by the pain and shock. He blames
the ground for being there, he blames everything around him;
his ability to cope suddenly dissolves. It is as if he were walking
a tightrope of development and even the smallest of pushes

sends him off balance and over the edge. Yet there are also those who tread a broader bridge as they grow. Nudge them and they may stumble, but they do not fall, or if they fall they manage to get back up quickly. *Vulnerability* and *invulnerability* have generated great interest in recent years because they cut through to the heart of an important issue. They are outstanding examples of early, predisposing tendencies that dramatically affect a child's response to the world. What exactly are these tendencies?

Perhaps we can see them best in a nice but admittedly simplified image of vulnerability and invulnerability described by the psychiatrist E. James Anthony as "the three dolls of Jacques May." One doll is made of glass, another of plastic, the third of steel. When struck by a hammer the first disintegrates, the second is permanently scarred, the last remains unscathed. These first and third dolls are the extremely vulnerable and the invulnerable. The problem with this image, as Anthony points out, is that there is no place for "the mechanisms of defense and coping since the dolls are inherently unable to develop reactive capacities in response to experience."[1] In fact, these very "reactive capacities" are part of what makes people function vulnerably or invulnerably to begin with. They are able or unable to adapt to their environments. Indeed, the predisposed level of adaptability is a key ingredient in much of what makes a child or adult vulnerable or invulnerable. It is a central component, but not the only one. We have come across others in earlier chapters: they are the building blocks of personality development, namely sensitivity, curiosity, engageability, human versus object orientation, and so on. And they arise through the hard-wired faculties of maturation we have also seen: perceptual, cognitive, and sensorimotor ability.

If it seems difficult to envision how an innate level of sensitivity serves as a precursor to vulnerability or invulnerability, consider it first in its earliest expression.

The Yale psychiatrist and researcher Donald J. Cohen notes that when subjected to natural stress, infants suck on pacifiers more vigorously (with fewer pauses between bursts of sucking) than they would normally, possibly as a way of modulating their distress. And when the stress increases above their threshold of endurance, the sucking falls off completely and the infants start to cry. When male infants two to five days old were monitored during their circumcisions, most stopped sucking immediately, began crying, and had extremely fast heart rates; they were overcome during the operation and for hours afterward. But interestingly enough, some infants continued to suck on their pacifiers, had only moderate increases in their heart rates, and were not especially irritable after the operation. In other words, some infants begin life less sensitive than others or with an ability to cope with displeasure, and although this and other abilities at birth may be influenced by conditions in the womb during pregnancy, genes set the stage. Although efforts by parents or teachers to help a child adapt to difficult situations can have enduring benefits, some children appear to be successful from birth – and they benefit from this ability the rest of their lives. Cohen writes, "These *persistent early adaptations* . . . shape the child's later adaptations and style of approach to new developmental tasks" (italics in original).[2] Some people, however, are not so successful.

The Vulnerable Ones

They wore their environment like an alien suit of clothes, with the unexpressed awareness that every suit of clothes is alien and provisional.

Max Frisch's line from his novel *I'm Not Stiller* captures the essence of vulnerable children – and the adults they become.

The environments of these people are seemingly alien and un-comforting. From birth on they have a hard time finding a state of serenity; they search constantly, unsuccessfully for inner peace. Nor can their parents find the right moderating care for them. Very little seems to help for more than a short while. Noises, lights, pain, disturbances of any kind push vulnerable children over the threshold between ease and discomfort. Average stimulation is irritating or alarming – or more subtly, average stimulation may be unable to reach such children and elicit appropriate responses. It seems that true comfort is always elusive to them, since they are disquieted despite every attempt to find harmonious conditions for living. Perhaps environmental deprivation or injury underlies cases like these. Their behavior may be similar to that of children under extreme duress. Yet these children are different: they are susceptible to injury even in normal, expected conditions of living.

One may look for the source of this frailty in different terrains of the mind. Since heightened sensitivity plays a part, as we have seen, there are possible organic difficulties as well as impairments in the maturational and developmental sequences of growth. And since part of the ego's function is to manage excessive sensory input and avoid unpleasant situations, we can question whether the ego apparatus itself is faulty, as it relates again to adaptability.

The following case reveals several of these aspects.

At the age of forty-eight, Laura is the only woman in a leadership position at a successful corporation. She earned it by devoted and tireless work. But she believes that her peers are superior to her, and that at any time she may be demoted; when the chairman of the board calls her to his office she always anticipates with great anxiety that she will be criticized and her weaknesses discovered.

She is divorced, but she protects her friendship with her ex-husband even though he is remarried and has new children. She is similarly devoted to her only son, who reacts to her affection with withdrawal and criticism. The only male friend she now has, someone who is deeply in love with her, she rejects because she feels he is boring and too inarticulate to share the intimate exchanges she desires. Thus, she sees herself locked into situations that yield little or no satisfaction or enjoyment, and her social shyness does not promise a more optimistic future.

In early life her younger brother suffered from many psychosomatic illnesses. He was severely incapacitated by a pain in his spine for which no organic cause could be found. She was sensitive to her body too, as well as to the way others related to it and to her: despite her early beauty she was usually excluded from groups because she was never able to assert her position or contest anything with her friends.

Her father was much older than her mother, marrying late in his life, and he continued his earlier habits of working hard and often overtime. Although he was a good provider and a caring husband and father, work was his priority. When the children were in their early school years he became ill and often needed emergency hospitalization, which cast a spell of uncertainty over the family. As a wife Laura's mother was servile to her husband's needs, and she delegated much of the care of their children to nurses. Thus Laura's mother became for her a model of nonconfrontation and servility; her father was the protector first and, when he became ill, the protected.

As time went on, Laura's lack of aggression and assertiveness revealed itself even more: nonconfronta-

tion became the rule for her. She spent a few years in a boarding school where she was acutely miserable, too shy to speak up in class and shy about her developing figure outside it.

Identifying with a passive mother, she could not enjoy her own active sexual urges, nor could she enter a significant relationship with a man. Her marriage lasted only a few years. She said she never knew what went wrong with it. Forever longing for continuity in relationships – and lacking the ability to please to achieve it – she was wounded by the breakup.

The sensitivities and low level of assertiveness that Laura showed from birth were combined with attitudes of her parents that channeled her into positions of even more discomfort and lower self-esteem. This intermeshing of both genetic and environmental influences became for her a lock she could not break. Because she was someone who could not extract from her parents the attention and care she needed, she grew up without it. And neither her professional successes, her attractiveness, nor a family of her own were integrated into a secure inner sense of selfhood.

Although environmental pressures are involved here, to view Laura's life history only in terms of external disappointment is to miss the inner forces at work from the start. By following the developmental path of identical twins reared apart we can see that at each stage where milestones are usually met, *both* vulnerable twins experience greater hardship than their peers.

An identical twin appeared irritable soon after birth. His pediatrician at first diagnosed his irritability as colic, later as an isolated symptom that would disappear in time. Yet the infant remained in a state of discomfort, was restless, and was unable to be calmed for very long. At seven months he showed an exaggerated stranger reaction; it had the intensity of panic. Later on, road construction in his neighborhood frightened him, as did other sudden noises. He formed the impression that life was dangerous and fear-provoking, and his fears turned to phobias. He avoided all situations that created in him a sense of alarm, and he later developed compulsive activities to organize and control his fears.

His twin brother, reared apart by more caring parents, exhibited similar symptoms.

Leave the invulnerable child to his own resources and he may steer himself clear of difficulties; but often as not the vulnerable one suffers, unable to fight free, and frequently blames the outside world for his discomfort. Although the burden of relief seems to fall entirely on parents, who may do all they can to repair, comfort, and soothe, their efforts are often thwarted. What they do, though helpful in its way, will not change the inclination these children share for adversity – an inability to adapt, high sensitivity, low engageability, and so forth. Doubts may soon emerge in parents' minds about their ability to nurture or love a child who seems so different from others, and the more they actively try to put him on a healthier path the more frustrated they may become.

In the case above, the mothers of the vulnerable twins had different attitudes toward their children. The first one gave up trying to soothe her son rather quickly, leaving him to join as best he could his many other brothers and sisters in their play and chores. He did not change, and the healthier children around him made his behavior more noticeable; yet he managed a lifelong friendship with an older brother.

The mother of the other twin felt herself to be a failure, assuming the burden of her only child's unhappiness and blaming herself for his weakness. Her decreasing self-esteem and her son's hypersensitivities made them an unhappy pair.

No one ever explained to the second mother that even the best mothering could not totally undo her child's fears. She felt burdened by her failed attempts, and the burden should have been eased. Still, she was right not to give up. Parents do have influence. They may not be able to eliminate predisposed vulnerabilities, but they can markedly alter their expression and at the very least aid in the way they are experienced. Attitudes ease or complicate every situation, and various approaches at modulating this child's discomfort may work much better than others. We have seen the stranger reaction, how children at the age of about seven months typically express their knowledge of what is known and unknown by turning away from any approaching, unrecognized person with fear or suspicion. This behavior is a healthy milestone in maturation, as we saw, for it signals an adaptive and growing awareness of the external world, a recognition of the boundaries that separate subject and object. But when adults approach a baby they typically come to feel rejected if they do not get a big smile and giggle in return for their effort. To make this normal situation worse, the vulnerable child has a more intense response to the stranger, and his reaction of discomfort turns to a panic that lasts far too long for the confused parents, who may find themselves encouraging the approaching person even more vigorously in order to give their baby a second chance at friendliness. The unknown adult steps close again, which needless to say only increases the baby's fear. At last the parents may feel they have no choice but to take the frightened child out of the room, as though withdrawal were the only answer.

Without recognizing that some susceptibilities in life are predisposed by nature, we will have difficulty understanding the vulnerable child. By accepting the possibility of constitutional influences, however, we begin to approach a child's individual inclinations with greater clarity, and we can then begin to nurture in ways appropriate to each child's condition. In the case of an excessive stranger reaction, for instance, no amount of immediate comforting may allay the child's fears. One solution, however – and one that recognizes the particular predisposition of this child – is to accept his distress for what it is, letting him observe the approaching person from a comfortable distance and in a neutral way until he can become known. The child may even become the active one, exploring his new environment under safe conditions as his own inner regulation allows.

The vulnerable child may also have more intense separation problems than his peers, and so a parent may have to be present, at least in the visual field, if separation at bedtime is continually difficult. The child's lower than normal threshold for discomfort will lead him to hold on to his parents more tightly, which should be accepted with patience rather than met by frustration. One's very attitude can help.

We might question parenthetically whether the sleeping arrangements to which modern Western nations have grown accustomed – that of shipping the child off to his own room as quickly after birth as possible – are really suitable for all infants and young children. There are few other species of mammal that separate the child from the family so early, and perhaps it is not surprising that childhood sleep disorders are on the rise. In any case, vulnerable children may need a nighttime parental presence for a longer period than other children, which does not necessarily mean that they have to share the parental bed; but it can include arrangements that allow for some kind of voice contact with mother or father.

As one can see, alternate methods of caring may indeed ease the burden of the vulnerable child, who can end up leading as fulfilled, talented, and accomplished a life as anyone else. Doing everything for this child, however, should not mean trying to mold him into a more familiar fit – which cannot happen. We have to stress again the significance of understanding. Recognition alone will neither eliminate our concern nor lessen our distress as parents, but it may reduce our sense of guilt, confusion, and helplessness. We can learn what can be done to soften or temper our reactions; we can learn not to preach in the expectation that force of will can overcome all difficulties. Throughout his childhood (and often into his adult life) we can convey a necessary sense of patience to the vulnerable child, esteem to foster his self-esteem. Though it will not be easy, this child more than others will need that extra care. And in addition to the positive influence it may have for him, it may instill in us a greater sense of what it means to be a parent.

The Invulnerable Child

If we can specify vulnerability as one of the "joints" of nature, a pattern of unwavering susceptibility to which the individual is inclined, then we may also understand its counterpart. Indeed, invulnerability is a pattern that has received a good bit of attention of late. Consider how a father describes one of his daughters:

> Beth was always happy as a child. She never seemed to ask very much in the way of attention, the way her sisters did. When her mother died she was very upset, but she's the one – of all of us – who's been able to

*pick herself up and get on with the things she wants
to do.*

We know this girl without having met her, for we have met
someone like her, or had someone like her in our family, or
perhaps read about her in novels or histories.

Some children seem to need relatively little from the
world in order to maintain development, and what they do
need they can extract from even limited resources. This is what
we mean by the term *invulnerability*. Its expression can be seen
in something so simple as a child falling on the playground,
scraping both knees, but running on again without crying. This
child is able to function under both the normally stressful and
abnormally difficult conditions of life.

> *A teen-age girl had to flee her town in Germany during
> the Second World War. She and her family traveled
> quickly, finding shelters each night until they were fi-
> nally caught by a small group of soldiers. She watched
> her parents die, and then her brother, and was spared
> herself only to become the subject of violent sexual
> abuse over a period of weeks, during which she wit-
> nessed the deaths of many people she knew. When she
> was rescued by Allied forces and then adopted by rel-
> atives after the war, she returned with rather unexpected
> speed to previous interests, avowing more gratitude to
> her adoptive parents than anger at her situation.*

This girl's curiosity for both everything new and for the meaning of her past suggests an enduring sense of survival, since
each of the traumatic situations she experienced was in itself
enough to have limited other people. This is not to say that the
images of atrocities, the mourning for her family, and the response to her uprooting did not deeply burden her, but in spite

of them, side by side, she proceeded to take from her new life what was offered and what she needed to continue living. Hers was an expression of startling adaptability.

How could she feel this way, living as she did with others who felt so much fear? Or put differently, how could she, feeling as much fear and uncertainty as anyone else, find the resources within her for survival, or like Anne Frank, even benevolence? What resistance to trauma did she possess that others do not? What thresholds for sensitivity, curiosity, adaptability?

Consider the children of alcoholics or drug users, or children whose parents divorce after years of open feuding, or those who have witnessed or are themselves suffering a prolonged illness, or had a death in the family, or live in squalid economic conditions, or have been the victims of sexually abusive parents. However severe their environments, some of these children seem to survive better than others. Examples of suffering and dislocation illustrate invulnerability in the extreme, but many children grow up under quite common conditions that partially or significantly subdue their siblings, that create anxieties, fears, or learning problems in others *but not in them*. As Yale child psychiatrist and researcher Albert J. Solnit depicts this situation succinctly: "Change always implies risk – there can be no psychological advance or improvement without risks. The question is whether the child can avoid or minimize the disadvantages of psychological risk, and with resiliency, take advantage of the opportunities for advancing development and increasing the capacity for mastery when change takes place."[3]

One recent study followed nearly seven hundred Hawaiian children raised in poor economic conditions, with parents who were often alcoholics or emotionally disturbed.[4] About 10 percent of these children managed to survive relatively unscathed, and some even showed progress. Researchers have tried to pinpoint the qualities that allowed these children and others like them to escape the effect of such difficult early environments. Susan L. Farber and Janice Egeland and co-workers suggest that such children have an earlier interest in their surroundings than more vulnerable children, that they have a relatively sturdy sense of tolerance, and that later in childhood they are better able to draw adults to them when they are needed and to form bonds. This "ability" can be labeled strength, activity, or assertiveness. But the term invulnerability suggests something even stronger – a higher threshold of sensitivity, a large degree of independence, and a secure inner regulation of development.

Explanations offered for this pattern of resilience (for instance, in the Hawaiian study) lie in the area of early protection – that some children who have had important or varied early attachments become less prone to stress and trauma later in life. There is a great deal of sense to this notion, and a number of researchers (John Bowlby, for one) have devoted themselves to understanding how patterns of early attachments set children off on their particular pathways of development. But this explanation, significant as it is, is not complete. It tends to drift back to a view of children as blank slates, *tabula rasa* organisms onto which experience alone draws its signature. It does not completely resolve the question of why some children from birth on are so much more able to have these satisfying early attachments than others.

An infant girl from a family in which both parents worked long hours somehow differentiated herself from her siblings and seized her parents' attention. By giving particularly strong signals when she was hungry or in need of stimulation, she demanded a response in a way the other children in her family could not. She imposed her presence and actively shaped her relationship with her parents. Simply put, she was assertive and could extract from others what she needed.

Admittedly, there are families who favor a particular child and heap on him an extra share of love, the way Joseph was favored by Jacob in the Bible. Such a favored child may be given a better opportunity to identify with good models; he may be better able to select from the environment because he himself was selected as favorite; he may, to borrow from Shakespeare, have "greatness thrust upon him." A child's environment may supply subtly encouraging cues that protect, inspire, and strengthen. Some children are sent such cues; others are not.

Yet some children seek out reinforcing signals more aggressively than their brothers and sisters from the very beginning of their lives. Children are not blank slates to be written on afresh; they are not equally resilient. Indeed, some bring to the world an ability to select what they intrinsically seem to know they need, and they then receive it with a full embrace.

We have to learn more about the healthy child and certainly more about resilience; less has been written about these children than about their weaker counterparts. The child who can help himself does not need as much from us, we often feel, and our energy as parents or therapists turns to the child who is troubled. But it is important to pay attention to these resilient ones, for they broaden our view of the interplay between predisposition and environment; they show us much about the factors influencing development, and much about coping with stress and trauma. Finally, since environment may have the final word on our survival – because no amount of invulnerability at birth can outlast starvation, a bullet, a bomb – such children reveal the power and limits of human resilience in a world that increasingly demands strength to survive in it.

Chapter Eight

The Individual "at Risk" in the Environment

I am amazed, methinks, and lose my way among the thorns and dangers of this world.
Shakespeare, *King John*

A lthough it is important to understand the inborn vulnerabilities or invulnerabilities we bring with us to the
world, it is also important to keep in mind how the
environment can put us at risk; for even a child or adult endowed with resilience can encounter forces in the world powerful enough to disrupt his development. And this notion of
risk will take us back naturally to the meaning of appropriate
nurturing.

There is an old, apparently true story about Frederick
II, who, born in 1194, succeeded to the throne of the Holy
Roman Empire in 1215. Despite the fact that he was renowned
as a patron of the arts and sciences, Frederick is also known
for a disastrous experiment with children. Assuming that there
must be a basic, ancestral tongue all people shared – most likely
Greek or Hebrew – he set about trying to discover what it was.
He took children too young to talk and ordered their parents
to care for them physically but not to speak to them. Without
hearing the spoken word, he assumed these children would
begin to speak all by themselves in what must be the true,
original language of mankind. But Frederick never found out
its name. Lacking stimulation and emotional nurturing, the
children soon died.

The environment puts people "at risk" when conditions
are so meager, attention so minimal and depriving, or so over-
whelming, that the likelihood of healthy growth is greatly diminished. Vulnerable people walk a tightrope of development
where even the softest nudge sends them off balance; invulnerable ones can suffer the same blows but manage to rebound.
Yet there are conditions in life no one can withstand. In a
society in which, as of this writing, over 13 percent of the
population fall below the poverty line – and in which half of
those households are headed by single women who must also
be wage earners – it is not difficult to imagine how the risks

for some children are increased. And of course the statistics worldwide, of infant mortality rates and poverty, are often astonishingly worse. When would-be healthy people are exposed to trauma – adults under continual bombardment in a foxhole or under torture who then lose their ego control, or children suffering constant physical and emotional abuse, or whole populations malnourished – any development can be derailed. Accept no argument of individual or racial or group weakness: the environment is powerful. It can end what any genetic plan begins.

History is brimming with examples of neglect. No less a humanist than Jean-Jacques Rousseau placed all five of his illegitimate children in orphanages. These "love children," as they were then called, lived in institutions not much better than the ones described in literature. Perhaps the workhouses of Dickensian England didn't kill the children who endured them, but to call even a day's residence there nurturing is to rewrite history. Extreme cases are easiest to find and hardest to look at, but of course there are also causes of emotional injury that have less to do with human catastrophe than with simply being out of tune with the needs of children. This is the more subtle but quite common path to anguish caused by insufficient or inappropriate nurturing.

The great challenge to parents and educators is more than to hope for the biological health of the child; it is to recognize individual needs and give sustenance to them. This notion is not new, just often forgotten. In "A Story for Aesop," John Berger writes that we might share "a tenderness for experience because we are human." Some people ignore it, yet some "fix the attention upon experience and thus on the need to redeem it from oblivion, to hold it tight in the dark." The care parents give their children is not merely influential; it is crucial to the healthy life of the child, dependent as he is. The first question, then, regarding how one relates an understanding of nature and nurture to child care, concerns the meaning of appropriate and inappropriate nurturing.

Although the general guidelines for child rearing are by now available in more how-to books than any parent can read, something is often missing in them – an awareness that care must also be tailored to fit each child's disposition. Appropriate nurturing must mean finding a way to rear that is suited to the child's own personality and needs, his own pace and variation of growth. As parents become sensitive to their children's un-folding maturation and development, they may then find them-selves in a better position to nurture in ways that are considered and specific. Children are not all the same – a fact everyone knows but often forgets when told how healthy children ought to behave and to be cared for – yet one can still make a number of general observations about how to provide for these individ-ual needs.

For instance, we have already seen the timed emergence of several human faculties that correspond to particular areas in the brain. Two of these faculties are especially important to the parent: affect (emotion) and cognition (thought/memory/language). Just as each child has his own sensory and motor apparatus, each expresses his unique affective and cognitive faculties and modes of behaving.

We know, first of all, that any regard for the child's development must be continuous and reliable. It is essential for the child to be able to expect events, to anticipate them, and so to form his own inner sense of regulation and security. When parents are either physically or emotionally absent, when there are frequent separations, when no single caregiver is reliably available, when illness in the household seriously limits child care, or when a parent is depressed – these and other situations can lead to affective or cognitive disorder.

The unwed mother of a two-month-old infant boy paid a next-door neighbor to care for her baby while she went to work. (The child's father had disappeared soon after the beginning of her pregnancy.) Although the neighbor was not physically abusive in any way, neither was she interested in talking to, playing with, or stimulating the child. In fact, she left the boy alone much of the day on the assumption that she would hear the child's cries through the wall if anything went really wrong. The boy gradually stopped making eye-to-eye contact with his mother when she came home after work. Six months later, frightened because her son appeared listless and unresponsive, she took him to a doctor, who, given the symptoms, suspected that he might be mentally retarded. Nonetheless he advised her to put the child in an infant care setting, and through the stimulation of caregivers this boy has begun to show signs of vigor and contact, though for the time being he is far behind his peers.

Very early on one can recognize the signs of children at risk. Their landmarks of development, such as early eye-to-eye contact, the social smile at two months of age, their exploration of the environment, the stranger-reaction at seven months, may be delayed or even absent. Lacking a reliable caregiver, the infant may turn to anyone for help, clinging to that person even though he may be in no better a position to provide continuous care. The infant who cannot search for or does not find and attach himself to a reliable caregiver will have a difficult time evolving the intimacy of an enduring relationship. For him, every person becomes unknown and interchangeable with every other, any lap to sit on as acceptable as any other.

Most infants do not wait passively to receive care and

stimulation; they seek it out. But if love is withheld and they are in a state of discomfort or longing for sustained periods of time, they will begin to feel helpless; their mood turns negative, their individuation and self-esteem disrupted. This emotional deprivation may later be experienced as a sense of not being wanted, not being recognized, or not being understood. As the child grows more unhappy and difficult, the parents may become more disappointed, with guilt and resentment replacing the earlier harmony.

If fundamental emotional contact is withheld, a child who is otherwise stimulated in such cognitive areas as language or thinking may still feel deprived. Conversely, if cognitive stimulation is withheld but emotional contact given, a quite different set of difficulties will arise. For instance:

> The eighteen-month-old daughter of young parents was handled, bathed, and dressed with regularity. Yet the parents demanded early discipline, imposing early feeding schedules and toilet training and restraining the child's motor activity with a halter. They believed their daughter should be "seen but not heard." Although she never missed a meal and her parents thought they were doing all they could to raise her, this girl was never given toys to play with – and until she entered nursery school she had never played with another child.

When parents don't provide toys, don't speak to a child in that often fabulous language they adopt or invent together, don't spark motor activity and perception through all the senses – in sum, when parents don't arouse the child in any way they run the risk of reducing brain function. They impede the natural timetables of growth because they don't give them a chance to unfold and flower. And hand in hand with this deprivation comes emotional retardation.

Interestingly, the overstimulated child may develop signs of disorders similar to those produced by insufficient care. He may appear in need of continual attention from parents and others, may want to be center stage at all times, or may appear shy, wanting to withdraw from the activity that tends to overload him. He may start to turn away from the primary caregiver who is constantly arousing. Such reactions are common when parents overactivate a child's skills or give him so many toys that none can be tested long enough to provoke curiosity and exploration. And some parents like to cuddle the child with such frequency or intensity that both he and his evolving independence become inhibited. By overgratifying the child's wishes, they restrain him from developing a tolerance for delay, without which he will always be bound to search for immediate satisfaction and be unable to plan for the future.

Moreover, there are situations, complex but not uncommon, in which deprivation alternates with overarousal, or when side by side, year after year, one area of function is excited while another is neglected.

Plainly, parents can do too little. They can also do too much, and it is also possible that they can do too little by trying to do too much – that is, they can give the disposition of the child too much sanctity and thus intentionally back away from regulating it. This is what has often happened in the wake of the reaction against the social restrictions of Freud's time. The lesson many people came away with was that repression, and the subsequent anxiety of uncovering one's sexual impulses, caused neuroses. They believed that saying no to children, no to the body, no to masturbation, and no to sexuality was an invitation to neurosis.

Many people, from the United States in particular, championed this cause, the cause of the new *yes*, and in reaction against repressive codes of moral conduct they went too far in their permissiveness, the effects of which are still very much felt. Permissiveness in child rearing has not resulted in a uniform social crisis, the kind that some people have looked for as proof that the older, stricter ways were better. But on the level of the individual it has in many cases proved less than successful. In giving children freedom parents often leave them alone, unguided and lacking specific stimulation and regulation in areas such as those we have mentioned, including sensorimotor abilities, language, thinking, and emotion – in sum, the regulation known as *continuity of care.*

Such laissez-faire rearing – simply satisfying the most basic needs of a child and then letting him go in whatever direction he wills – fails because it does not recognize how intertwined nature and nurture really are. Biological inclination is not geared to churn out a preformed adult but rather to respond to the environment throughout development. That with which we are born needs, one can almost say seeks, outside nurturing and stimulation in order to survive. Since our innate flexibility allows us an individual range of response, environment should not be imagined as the second half of a one-two punch (biology being the first) that determines life. It is rather an integral part of the process that *is* life. We need and seek regulation from outside.

Recognizing individual disposition means understanding what is appropriate for the needs of a particular person at particular times in development – needs for privacy and solitude, needs for help and regulation. Some people require rules and schedules at certain times, whereas others find these same demands too restrictive at any time.

Finding the right measure as a parent or teacher – not doing too little or too much, maintaining regularity without being too strict – can be extraordinarily difficult. Success often depends on one's ability to recognize the child as someone with a history and disposition rooted beneath the variations of the moment. If one really knows the child, one knows him through the changes in life that not only make his personality but also reveal it. Being able to support this individual throughout development with whatever resources one has available means being able to detect his flow and his patterns – the patterns that reveal how he adapts to life.

Inappropriate care can have severe consequences. The conclusion jumps out at us; never veiled, just occasionally forgotten: environment has the last word. It can stifle any endowment. The question we must always seek to answer is how environmental care can best nurture the nature within. The demands of nurture are no less important than those of nature, for "life," as Erikson says, "does not make any sense without interdependence."

PART III

*Nature and Nurture
Writ Large*

CHAPTER NINE

A Long History
Briefly Told

What's the use of learning that I am one of a long
row only – finding out that there is set down in some
old book somebody just like me, and to know that
I shall only act her part; making me sad, that's all.
The best is not to remember that your nature and
your past doings have been just like thousands' and
thousands', and that your coming life and doings'll
be like thousands' and thousands'.

THOMAS HARDY, *TESS OF THE D'URBERVILLES*

P eople usually prefer to think like Hardy's Tess Durbey-
field: if the past seems a burden, to ignore it is liberating.
To ignore it means to free ourselves to discover the world
outside, the world of choice, of experience, of possibility: it
means to be free in a future in which we can become who we
want to be. Haunted by her past, Tess believes that "true history
lay, not among things done, but among things willed."

To the degree that everyone can follow Tess's hope and
ignore the boundaries defined by the past, and to the degree
that society treats us all fairly in this venture, we can indeed
become whoever we want, bound only by the limits of our
desire. But if the freedom to be oneself in a fairer world is
difficult, the freedom to be someone without a past at all is
simply impossible. What is neglected or forgotten in wishing
for unlimited personal flexibility – and what Thomas Hardy's
novel is full of – is the part of us nature helps to define: our
predisposition, our hereditary past, the core equipment that
makes us as different from others, or alike if we're identical
twins, as the lives we subsequently experience. We have explored how heredity inclines each individual to a course of
maturation and development, influencing everything from ru-
dimentary perceptual abilities to temperament, psychic defense
mechanisms, and adaptation. It helps orient each of us to the
world in a unique way.

The United States is perceived as a land of opportunity,
and restriction is no more part of our national mythology than
it is of Tess' personal one. Despite a history of slavery and
continued racism (or perhaps because of it), we struggle with
a national mythology in which no one should ever be bound
by the conditions of his birth, that everyone should live with
the possibility of betterment, and that, as for Tess, "true history
lay . . . among things willed." Many of the immigrants who
have come to America harbored the hope that they could work

hard, establish homes, and raise children who would be better off than they.

As a result, we have a rather strange relationship to the concept of the gene. On the one hand there is an anxiety over genetics that has deep and painful roots, among them a shared memory of the Nazi policy of eugenics in the 1930s and 1940s – of efforts to improve the gene pool by eliminating "undesirable" carriers (in this case mostly Jews and Gypsies but homosexuals and the mentally ill as well, among others) – that stays vivid in our minds. Heredity has been used as an excuse for prejudice, for keeping whole groups subordinate to others with the rationalization, conscious or otherwise, that "nature" so intended. The notion of heredity has been ill used in a variety of ways, and perhaps for this reason genes have taken on the negative role of determining life before we have even had a chance to live it. For if we feel genes really organize every aspect of who we are, then what chance can we possibly have to improve ourselves, to make our own changes, to construct a future in which we are entitled to live happily? Genetic determinism, the genetics of groups, and the eugenics practiced by such groups as the Nazis are ideas that still cast a long, bitter shadow over the word *gene*. Hidden prejudice, if not overt racism, lies near them; freedom slips away. From this perspective, the gene is intractable, unmerciful: it is a "Big Brother" watching us not from above but from within; and people are wary of giving up such control.

On the other hand, however, the science of genetics excites us. In newspapers recently we see the headlines of a new science: MAJOR PERSONALITY STUDY FINDS THAT TRAITS ARE MOSTLY INHERITED; BURST OF DISCOVERIES REVEALS GENETIC BASIS FOR MANY DISEASES; SCIENTISTS "MAPPING" CHROMOSOMES; POTENT TOOL FASHIONED TO PROBE INHERITED ILLS; GENETIC ENGINEERS PREPARE TO CREATE BRAND NEW

PROTEIN. Despite certain hesitations, those who might benefit from the study of the human gene and those who simply delight in the wonder of its potential – just most of us – are also in awe.

The conquest of disease, one of genetic science's most dramatic contributions, amazes us in a way, perhaps as we are amazed when hearing of moon landings and superconductors. Achievements of this kind appeal to our admiration for the unlimited power of science. Yet we stay wary of the reputation that genetic determinism seems to possess. We become passive recipients of the news; the discoveries are events that seemingly happen *to* us and over which we have no control. And one result of our passivity is cautiousness.

Future historians will look back on our time as one filled with an explosion of enthusiasm for the study of the gene and the solutions it may bring, and also on our hesitation to accept it. How did we come to such a collision of moods, to such a rich ambivalence?

In traditions such as those of the East, life's opposites, like yin and yang, exist simultaneously as part of living; contrariety is possible, its nuances even esteemed since they reflect the whole. But we in the West – to make a broad distinction – have a tradition of making broad distinctions. We reduce unities into elements and elements into subparticles, never to be rejoined as part of the whole from which they came. The debate over heredity versus environment is an example of this reductionism. Writers, philosophers, politicians, and scientists have decided for themselves over the centuries how and where to separate them, and which to champion over the other. Fractured into separate pieces, like a rock split with a wedge and a sledgehammer, these influences came to be seen as warring opposites, with one more powerful than the other depending on whom you ask and when you ask it. But there is a danger in so much splitting, that it is done too hastily and with bad aim: nature's joints are not carved but rather fractured into ideologies.

The trail of these fractures lies in a history we can follow. However, there are really two histories, not one. There is the recorded legacy of the writers, philosophers, politicians, and scientists who attempted to make sense of the world as they saw it; and there is also the mostly unwritten sensibility of the people living during the times of these thinkers. The first history recounts the battle among giants over the realm of ideas. The second is the living understanding of people's lives that goes on unnoticed beneath the province of ideas. These two histories are not separate, of course; each influences the other. But to understand how the popular conception of human development came to be what it is today, with its emphasis on the role of experience and learning over that of heredity even while new information about the gene explodes all around us, one has to follow the readers in history as much as the writers.

Long before James Watson and Francis Crick's discovery of the structure of the DNA molecule in 1953, a person's nature was seen as his destiny, his inherent "natures" and "natural states" the very stuff of which he was made. By rough count, there are twenty-four references to "nature" in Shakespeare's *Macbeth,* and many of the twenty-four refer directly to the play's major characters and the character of the characters. There are references to "villainies of nature" (I, 2); "wild in nature" (II, 4); "royalty of nature" and "patience so predominant in your nature" (III, I); "a good and virtuous nature" and "intemperance in nature" (IV, 3). In sum, "every one according to the gift which bounteous nature hath in him closed" (III, I).

As we mentioned already, Thomas Hardy, the poet-explorer of the distance between natural law and society's law, wrote (despite Tess's hopes) of a wide array of the traits he considered inborn. Within a thirteen-page span of *Tess of the*

D'Urbervilles, for instance, there are references to "impression-able natures," "physical nature," "imaginative" natures, and "natural shyness." Later we see Tess's "rougher nature," and her "natural attractiveness"; plus, she doesn't "believe in any-thing super-natural," such as society and religion.

Of course, simply because someone thinks, says, or takes the time to write that so-and-so possesses "royalty of nature" or a "natural shyness" neither means that "royalty" nor shyness per se are necessarily inherited, nor even that such qualities in other forms – confidence/grace or humility/modesty – are genetically linked. But to label these early attempts to define nature's boundaries as simplistic is to miss the point. What is interesting is that long before the work of Watson and Crick, or even before people became aware of Mendel's dis-covery of the laws of hereditary transmission in the 1840s, thoughtful people did have a notion that part of what makes us individual is passed on through generations. It is a sensibility that seemed to carry a seed of sense.

However, John Locke in the seventeenth and Jean-Jacques Rousseau in the eighteenth century each expressed a classic vision of environmental influence that was so forceful on both the political and the philosophical planes that we to-day are still moved by their influence.

John Locke is remembered for his empiricism, his dis-avowal of the rationalist belief in innate ideas, his later influence on Hume, Berkeley, and other Enlightenment thinkers, and his role in developing the concept of political freedoms. He was also the first to articulate a pure "nurture" position. In such a position as Locke's, nurture stands in opposition to nature. It mends and molds. Nurture is poet to nature's sergeant. It is experience, possibility, flexibility, and education. In his *Essay Concerning Human Understanding* (1690) he writes, "No man's knowledge here can go beyond his experience." All events and learning after birth are experienced in the mind as though drawn on a blank slate, or tabula rasa, where they accumulate as knowledge. There is no room in this empiricist vision for innate ideas, and thus no room for natural human states of bleak conflict and greed such as Hobbes described. For Locke, the only natural state of man is equality, in which all people have the right to pursue life, liberty, and possessions – and his list of basic freedoms was co-opted nicely by the founding fathers of the United States. Jean-Jacques Rousseau's *Discourse on the Origin and Foundations of Inequalities Among Men* (1754) echoed these themes, and perhaps for the first time the idea of the importance of a general education for all was given its due.

Still, what average guild workers or servants in England and France actually thought of these liberating notions at the time, if anything, is harder to know, and in a sense their understanding is just as important. For instance, Locke's notions must have been at odds with the rigid class boundaries and the power of lineage in contemporary British society. If the family to which a person was born determined the dominant course of his life, from class to profession to range of potential mate, what could be so meaningful about life's learning experiences? If the understanding of heredity at that time was of the sperm as a miniature adult, deposited in seed form into the female by the male for a few months of warming up before it came out as a little person, or if the laws of primogeniture ensured that the firstborn son would inherit all the land of the father, or if, despite the existence of a parliamentary system of government, there was still a monarch to produce future monarchs and a ruling hierarchy of lords and knights, then what chance at betterment truly existed for most people?

England was not the only country about which these questions can be raised; how much spirit Rousseau actually breathed into every household in France is simply unclear. Even though France and the American colonies brewed popular revolutions within a decade and a half of each other, both hailing equality as the natural state of mankind, the question of whether equality extended beyond one's friendly neighbors to a place like Africa, or to Gypsies or Jews, was a different matter entirely. The Bible's "Love thy neighbor as thyself," which in its essence reflects the unity and mutual destiny of all people, unfortunately left open the definition of *neighbor*. Were Africans also neighbors? Even though Locke and Rousseau highlighted a new age of ideas about freedom, a subterranean river of status quo (us-better-than-them) thinking no doubt also stayed alive.

For this reason, although older notions of hereditary lineage may have been fading in some ways as the so-called middle class was growing (giving every family some chance at comfort), at the same time even the most liberal thinkers were capable of prejudice toward those they perceived as different. The fact that slavery existed in the United States for nearly a hundred years after the signing of the Declaration of Independence, prompting a devastating civil war to end it, gives some idea of the depth of racial bias even within a climate of new freedom. Whatever his other attributes, Thomas Jefferson himself owned slaves.

We get an even better idea of this old-fashioned bias in the social Darwinists of the late nineteenth century. Charles Darwin published *The Origin of Species* in 1859, and it did not take long for people to seize on a misunderstanding of this work for their own ends. They had their reasons, of course. The rise of the middle class put the children of average tradesmen closer than ever before to elite schools, better neighborhoods, the government. The threat was not from African blacks, over whom the upper classes could claim racial superiority and justify their colonial involvement; this threat was from approaching, like-faced neighbors, so from the upper-class point of view it was best to put class distinction in new perspective. Many conveniently found this perspective in Darwin: Thus, (they said), the rich sit where they sit for natural reasons. The survival of the fittest meant the superiority of the rich and powerful over the middle-class and poor; after all, would the rich be rich if they weren't naturally superior? It stood to reason that they must be fitter than the lower strata of society.

Such acrobatics of logic and misuse of the concept of heredity were also applied especially well against minorities. By the late nineteenth century every possible explanation was strummed up to prove the natural inferiority of people from the Southern hemisphere – from cranial sizes and lower innate ability on so-called intelligence tests to observations of general laziness. Politics and pseudoscience have a bad tradition of mixing, and those advancing politics in the name of such science fought the idea that every person has an equal chance to improve himself and the conditions of his life.

The dust clouds that rose from the battles over these issues have never settled. Equality versus racism became de facto choices for many people politically, and nature and nurture themselves became polarized and politicized. Nature came to suggest racial distinctions, immobility, and stereotype, whereas nurture implied education, improvement, and equality. And these battle lines were drawn in red colors.

Mass extermination along ethnic lines was not unknown before Hitler (consider the Armenian massacre early in this century), but the Aryan supremacists developed a more systematic structure and vitality to their program of genocide. If the surviving world population needed a better reason to bury racism and the bias associated with heredity, it surely could not have invented one.

At the same time, the "liberating" efforts to form societies of truly equal co-workers were proving unjust in their own way. In the twentieth century, Marxist and communist revolutions created nations in which environmental change itself was thought to remake man's nature, for by changing the economy and shifting the means of production to the state, greediness and competitiveness would be eliminated; everyone would contribute equally to a society of equals. The millions who died or were imprisoned under Stalin's rule in the name of this dream may obscure the genuine hope people had in it. In addition, because empirical science could not provide a rationale for a noncompetitive society, the Lamarckian theory of evolution was adopted as truth by Russian scientists. Lamarck's belief that parents could pass on acquired characteristics to their offspring – for example, that a blacksmith who developed large muscles could transmit his muscle bulk to his children at birth – upheld contemporary Soviet ideology that change was ever possible. But it didn't do much for science.

Science.

History has a way of concretizing ideas that ought to have remained plastic, and "nature versus nurture" is one of these ideas. It may be convenient for people to have large umbrella categories as receptacles for testing theories, but these particular categories, nature and nurture, deceive. Possibly because of their complexity, possibly because of the way "final" answers have been advanced so quickly and firmly on both sides, possibly because of the emotions involved, the old battle is less than useful now – because nature and nurture are not in opposition. They are, in all the ways we have explored, in specific interaction.

This interaction, however, is still by no means accepted, and we don't have to look further than B. F. Skinner for a recent example. Following Pavlov, J. B. Watson's famous challenge provided a goal for Skinner:

> Give me a dozen healthy infants, well formed, and my own special world to bring them up in, and I'll guarantee to take any one at random and train him to

become any type of specialist I might select – doctor, lawyer, artist, merchant-chief, and yes, even beggar and thief, regardless of his talents, penchants, tendencies, abilities, vocations, and race of his ancestors.[1]

The fact that Watson spoke at all of "talents, penchants, tendencies, abilities" in making his proposal undercut a purely environmental position, for it explicitly recognized innate predispositions, and thus heredity, as part of each person's makeup. But that important qualification, even if made reluctantly, was never conceded by B. F. Skinner.

In the 1950s and 1960s Skinner brought the social-scientific field of behaviorism fully into focus. If behavior could be manipulated by a series of positive and negative reinforcements, which he tried to show in his now famous experiments with pigeons and rats, then why look deeper than learning for an explanation of human action? Why look to, say, psychological motivation? Skinner's work was thought to be a direct assault on Freudian theory, stereotypically depicted as attributing all adult behavior to patterns of drives and defenses organized in early childhood. What settled the matter against Skinner, however, was ultimately not the strength of Freud's theory but rather a series of experiments revealing the species-specific behavior of animals. These experiments proved that species have their own innate range of behaviors and that general laws of learning do not apply to all animals, including human beings. You can train a rat for a decade but he still won't learn to make certain visual associations that a person or even a pigeon will. Learning models must take nature into effect.

Still, Skinner's learning theory while it lasted rode a tide that was felt beyond academic circles alone; it shaped the zeitgeist, the spirit of the age, and coincided not only with the popular acceptance of Benjamin Spock's feeding schedules for infants but with a growing disavowal of racist theories, with the accumulating if belated realization of the horrors of the Holocaust, and with the rise of the American civil rights movement. Once these social issues became embedded in public consciousness, the role of genes became relegated to determining obvious, narrow dimensions: certain diseases, facial resemblance among relatives, body size, eye color, blood type, and occasionally specific arenas such as musical or mathematical talent. Genetic makeup and social optimism came to be seen as angry opponents. With Skinner on one side and on the other the confused biological determinism of someone like Shockley, who confounds genes with intelligence and race, it's no wonder that a lingering confusion surrounds heredity's very real interaction with experience.

Yet as scientific research uncovers the ground plans of growth and their interplay with the environment, it may rescue the gene in the public's mind from its near fatal embrace with racism. Genes have been used as an argument between races, but they themselves are not racist. They are DNA codes. And in the last thirty years they have been found to influence much more than physical features.

We have reached a moment in our history when scientific research can take new steps. Genetic biotechnology will tell us much in the next few years, certainly in the next dozen, and it will, if we listen, change our perception about the meaning of genes in our lives. Politics must be left out. Ideology may take us away from ourselves for a time, but in the end a fresh and fair approach to understanding both our heredity and the environment that cradles it must be forged. When this happens, we will definitely have come a long way.

CHAPTER TEN

Flexibility in an Ordered World

Genes set limits to ranges; they do not provide blueprints for exact replicas.
STEPHEN JAY GOULD, *THE MISMEASURE OF MAN*

W hat has evolution bestowed on our species, incredible order or unflagging flexibility? In this age of technological breakthroughs in genetics, molecular biologists have been forced to consider this question in the largest and the smallest organic units side by side, with no great certainty as to which exactly are the largest and which the smallest. The cell was once the smallest known organic unit, but that was a long time ago. The nucleus had its day, way back when. DNA was a discovery, then amino acid chains, then particular amino acids, such as one called glycine. If one glycine amino acid is altered amid the 333 that are themselves among the total of 1,000 amino acids in the molecule that makes up human collagen, the result is a disease called *osteogenesis imperfecta,* which kills at birth.

Large questions arose concerning bodies and species, but they too have blossomed beyond their earlier borders so that one must now speak, as one speaks about glycine, also about society and evolution, about ethics, about morality – about, if one wishes, God.

In this chapter we move from large to small and small to large, as present knowledge allows, keeping in mind that no border is secure, no subject either small or large enough to satisfy anyone's curiosity about evolution a hundred years from now.

On the small end of things human first, there are about a million million cells in the human body, and each cell has a nucleus. Within each nucleus, in turn, are the identical twenty-three pairs of chromosomes, forty-six in all, and each of these chromosomes comprises thousands of genes, which are proteins. Each protein strand is a sequence of usually a hundred or more amino acids, which come in twenty different varieties depending on how three base chemicals are joined.

Together these chemical-based amino acids form part of the molecule known as DNA. When the protein strands split and replicate, new DNA is formed. The new DNA is an exact replica of the old. And from each of the twenty-three pairs of human chromosomes, some genes send signals in the form of new proteins either to other genes, turning them on or off, or to the cells they occupy. The signals are instructions. Our bodies live and breathe by these instructions.

The same is true for smaller bodies. For instance: there are exactly 959 cells in the adult *C. elegans* worm. After five years of patient study, researchers finally have a good feeling for how each cell in this worm gets to be where it is and do what it does – muscle tissue cells, digestive tissue cells, nerve cells, skin cells, and so on. Five years for 959 cells. Imagine the task of analyzing each of a million million human cells. How many years for that?

In addition, the skin of the *C. elegans* worm is transparent, which makes its examination all the easier. If one were to take a single human organ such as the brain, one would be faced with tens of billions of cells covered by a non-transparent skull. And there are whole sections of ill-defined nerve tissue in the brain of which scientists have no comprehensive working knowledge, not to mention of individual cells, not to mention individual molecules of DNA within those cells. It has been estimated that we have an understanding of only 10 percent of DNA in the human genome. We simply don't fully know what the other 90 percent do. So scientists often return to the worm.

One of the interesting things about *C. elegans* is that during its early embryonic stages it can be mashed up quite drastically and will order itself into proper shape again. The cells do not care where they are positioned, up or down; the proper wormlike shape of the organism will ultimately appear. But when the worm becomes an adult it is no longer an amorphous mass. Worms have various kinds of tissue and a nervous system and a digestive system and in many ways are quite similar to, say, mice. If cancer cells are injected into a mouse embryo early enough in embryonic development the cells will be transformed into working, healthy cells. The malignancy will disappear. If the same injection is made after a certain point in the development of the mouse, however, it will be born with the cancer and die. Similarly, after a point in the development of the worm, if you mash it up, you'll kill it. Ask any child who's ever gone digging in a garden.

It is at this point in embryonic development that the cells, which were once flexible, become differentiated in function. They become nerve cells, for instance, and chunks of them just won't work in the digestive tract if you put them there. Once cells specialize they become more or less fixed in function, but an additional step is also necessary. Many cells migrate as the worm takes shape in development. At a given time and by a certain trigger, these cells start communicating with each other and attracting one another into position. Just as one can identify a group of Boy Scouts in a crowd by the color of their caps, markers on a cell's surface membrane are read by surrounding cells, which attract their like-markered brothers. The exact process is by no means well understood, but one thing is certain; the process is a chemical one, and by chemical we also mean genetic.

The genes of each cell are responsible for differentiation and specificity. They are coded to turn on and off at set times and by communication with other cells. The pattern of the whole organism, in fact, is preset. The worm will have a worm shape and act worm-like always. It is an incredible order.

What is it that directs the overall movement of cells into their familiar shape and appropriate function time and time again? This concept is one of the harder problems geneticists as well as cellular and molecular biologists have tried to solve. One answer may be found in a cluster of genes, a set of master control genes that coordinates the dominant movement and development of each organism's cells. This set of genes, known as a homeobox, has been found in the chromosomes of *C. elegans* as well as in human chromosomes. Its existence implies that the structure of worm or mouse or frog or child does not arise bit by bit or mysteriously, but through an underlying integrity controlled by the genes and, presumably, selected for over hundreds of thousands of years of evolution. Noses do not grow on toes, fingers do not appear on ears. Creatures whose noses did grow on their toes by mutation probably did not reproduce as often, if ever. They did not survive. The bodily arrangement we share with other members of our species, and even with different species, is one that has worked well in the sense that it has enabled us to survive and reproduce.

The homeobox, the genes regulating the switching on and off of other genes in development, ensures regularity; it ensures similarity. Earlier we shied away from discussing the human brain, but we can now say this much at least. The structure of all the tens of billions of human brain cells is determined by our genes. God may or may not have a place in the universe, but the grand mystery of our shape, of how it could be that we walk around on two legs, and have arms of equal length and eyes not too far apart or too close together, and male or female genitalia, and a dual-hemisphere brain – these things are not a mystery anymore. They are genetic. "The overall evolutionary transition," writes Melvin Konner, "requiring millions of years, is fundamentally, relentlessly genetic."[1]

Knowing that even bodies of great complexity are ordered by genes allows us to answer many questions about such bodies. When we consider minds, personalities, behaviors, cultures, and societies, however, we move into a sphere involving larger questions. The solid information about bodies cannot be applied directly to dimensions such as social behavior without hesitation. What may be a good answer at one level – for instance, that of the gene – may be inappropriate, or deadly inaccurate, at another.

Genetic determinism, the idea that all aspects of life, not just bodily structure but also all behavior and culture, are prearranged by nature, makes this leap too freely. One must admit that the argument has a certain appeal: if the brain has evolved through natural selection and genetic changes, then consciousness itself, which is afterall a product of the brain, must also be genetic, as must intelligence and ability and character and finally even mankind's external expression of itself, namely culture and society. Genes do play a vital role in dimensions such as temperament, perception, emotion, and both physical and psychological illness (not only at birth, as we have seen, but through the life span); but genetic determinism reaches beyond these areas to include all of society's structure and functions.

This is a precarious leap. As mentioned in the last chapter, it opens the door for pseudosciences such as social Darwinism. Yet genetic determinism is wrong, that is to say scientifically 'incorrect,' for two other reasons as well.

The first reason involves evolution. Natural selection acts on organisms through their success in surviving and in reproducing themselves. Conventional Darwinian theory has it that natural selection acts not on each isolated trait of an organism but on the organism as a whole; each trait, however, serves an adaptive purpose and has been "selected for" over time. For example, fair-haired people prospered better in northern climates than dark-haired ones, perhaps because the sun is a source of Vitamin D and those who possessed slightly lighter hair and skin could absorb the rays better, were consequently healthier, and survived and reproduced in greater numbers, bringing the genes for fair hair together more often. Genes for fair hair ultimately became part of the genetic arsenal of northern peoples. It is a trait with a history. We might imagine that in the course of, say, four million years of human evolution very few extraneous or useless traits have become part of our bodies' genetic package. Everything that's there is believed to have some function that would help the organism thrive since it has been through the process of natural selection.

Interestingly enough, however, there are problems with this neat evolutionary pathway. Recent discoveries have indicated that a great randomness characterizes survival. Mass extinctions have eliminated a great percentage of all living species, and even the eventual development of *Homo sapiens* was anything but preordained. We are, it seems, quite lucky to be alive. In addition, and more important, many human features serve purposes secondary to the primary ones for which they were selected, purposes one could say that "nature" never intended. For instance, although we will one day soon, through the mapping of the chromosomes, come to prove the function of most of our genome, there are whole sequences of DNA that repeat and repeat themselves *without* an apparent function. Stephen Jay Gould, a paleontologist and historian of science at Harvard University, discusses these sequences of DNA in the context of a larger argument against faulty assumptions about the logic of adaptation. He writes that simply because these DNA sequences exist in the body does not necessarily mean that they serve an adaptive purpose. They may exist independently of the body's selection, replicating themselves simply because they can, in just enough quantity not to interfere with anything genetically important.[2] They are like remoras living off a shark, silent but ever present.

Whether geneticists will one day be able to specify their use, or whether these so-called selfish DNA are truly just along for the ride, useless but benign, remains to be seen. The point is that not everything in the body, theoretically at least, exists by grace of a long evolutionary process tied directly to control genes that regulate function.

The essence of the same idea permeates a number of other interesting essays by Gould. In "Of Kiwi Eggs and the Liberty Bell," for example, the question is raised why the Kiwi bird lays eggs so large that they could belong to a bird six times its size. The kiwi's egg, Gould believes, did not grow large because its size conferred advantages to the kiwi in its struggle for survival, which would be the strict Darwinian view. Rather, it is possible that some traits function in ways for which they were never selected. To look for adaptive reasons why the kiwi's egg grew so big is to look at the riddle from the wrong end. Evidence suggests that the kiwi's egg seems large for the size of the bird only because, in the course of evolution, the kiwi bird became smaller while the egg remained the same size. The egg did not grow; the bird shrank. Gould writes:

> I am satisfied that kiwis do right well by and with their large eggs. But can we conclude that the outsized egg was built by natural selection in the light of these benefits? This assumption – the easy slide from current function to reason for origin – is, to my mind, the most serious and widespread fallacy of my profession, for it lies embedded in hundreds of conventional tales about pathways of evolution. I like to identify this error of reasoning with a phrase that ought to become a motto: current utility may not be equated with historic origin, or, when you demonstrate that something works well, you have not solved the problem of how, when, or why it arose.[3]

Like the kiwi egg, not every physical feature has a history of natural selection behind it, nor do all features of personality and behavior. But first, we said there were two reasons why genetic determinism was untenable. Here is the second: the power of the environment.

Genes are inextricably involved in every nuance of the brain's embryonic development, sending signals to each other to specify the location and function of cells and producing other proteins that spur every hormone and enzyme. Although each feature may or may not have been crucial to the survival of the organism – been selected for and become part of the functioning genome – each human feature has been molded by living. The power of environment mediates, manages, and alters what genes predispose. And it acts on many levels.

Consider environment on the cellular level. What can happen when the conditions during embryonic development are harsh? Taken by mothers during pregnancy, a drug thought to be a sedative (thalidomide) resulted in the birth of infants with deformed limbs or no limbs at all. On one side are millions of years of evolution and genes that specifically read and control the growth of every part of the body; on the other is a bottle of pills that completely derails nature.

Another example. Newborn rats were separated into two groups and given different amounts of stimulation during development. The lesser-stimulated rats were literally locked away alone in dark cages, and when their brain tissues were examined, they revealed a poorer network of nerve cells in the visual parts of their brains than did the stimulated ones. Experience changed biological structure, which should come as no surprise. Good or bad nutrition changes bodies for better or worse every day.

In a very simple way, we all know that the environment affects us. It can have the last word. Any bullet can end what nature begins. The question is not whether environment has an influence. We know it does, always. A better question is how and when offerings from the environment interact with the demands brought by our genes and our individual biologies. We have seen that the genes, which send signals and codes for proteins that make nerve cells and enzymes and hormones that then regulate all bodily functions, actually depend for their survival on the body's adaptation to the environment. Adaptability is an integral, built-in part of biology.

Against an early tide of scientific criticism, the work of the molecular biologist Barbara McClintock has shown that DNA, the virtual cornerstone of life, is transposable, meaning that parts of a DNA strand may be picked up and read by other parts at various times. DNA is therefore somewhat malleable. As Evelyn Fox Keller puts it, McClintock demonstrated that DNA is "subject to rearrangement and, by implication, to reprogramming. . . . Such reorganization could be induced by signals external to the DNA – from the cell, the organism, the environment." One implication of this DNA transposition is interesting. Transposition "allows for the possibility of environmentally induced and genetically transmitted change."[4]

We have seen how any cell in the early stages of embryonic development is incredibly flexible. It can be put here or there or have cancerous cells injected around it and still develop normally. Staying adjustable and open to change would appear to be advantageous to survival. If it's true that "flexibility is the hallmark of human evolution,"[5] it is reasonable to believe that a certain range of flexibility is built into our programs. Creatures with the flexibility to survive in varying conditions would be preferred. With too much or not enough nutrition, for instance, we do not always or quickly die. Our bodies can conform to circumstances – but only to a certain degree. There is a range of adaptability built into our biologies, a range encoded in our genes.

It is safe to say that in a decade or two at the most, much more will be understood about genes, their links to disease, and how they underlie areas such as complex behavior.

New knowledge will dwarf the old. As genes are mapped and their chemical changes identified, the sequences for any number of traits will be pinpointed. This work is already underway.

What we will look for ultimately, however, is more than a decoding of individual features, for the lens of flexibility finally demands a focus on growth and adaptation. Knowing that a feature has genetic roots is not the same as knowing how it is expressed, why it appears differently in different people, and what makes it emerge and recede in the course of a day or a year or a lifetime. Knowing its biological underpinnings does not tell us everything about its later expression. Although the chemicals they fire are precise, genes express a range of possibility that is often imprecise, especially when it comes to human behavior.

This is why any attempt to link a specific gene to a specific behavior is extremely tenuous, and even a cluster of genes may not be a good predictor of actual adult behavior. What we do know is that our genetic makeup predisposes each of us to individual variations of the fundamental building blocks of temperament and growth that in turn incline us to individual patterns of maturation and development, with subsequent adaptation distinctly our own. The likelihood of predicting an adult's complex behavior based on a knowledge of his genes will only improve when we can decode the genes governing his individual rate of growth and his range of response to the world.

We express order amid variety, and variety amid order, and both have stood us in good stead. Efforts to deemphasize either the specific order behind our variety, or the fantastic individual variety behind our order fall short of the mark.

What, finally, of the most complex social behaviors, such as altruism and religion?

Erik Erikson writes that "each successive stage and crisis [in development] has a special relation to one of the basic elements of society, and this for the simple reason that the human life cycle and man's institutions have evolved together." Yet sociobiologists like Harvard's Edward O. Wilson have tried to link genetic makeup to all forms of social behavior. They say, for instance, that people are altruistic because genes for altruism benefit the group's survival and are selected for in the course of mankind's evolution:

> Facts are in accord with the hypothesis that human social behavior rests on a genetic foundation – that human behavior is, to be more precise, organized by some genes that are shared with closely related species and others that are unique to the human species."[6]

Sociobiology is not the same as genetic determinism, and it does not rationalize racism; but it does tie a shorter leash between genes and behavior than some believe. Gould's argument against assigning an adaptive history to every bodily feature must be heeded. For example, a sociobiologist might say that religion arises in society because of mankind's need for tribal sharing and cohesion, which increases the likelihood of survival and has come to be directed by genes. But what about religion as a response to the knowledge of our own death? Gould proposes that although man's knowledge of his mortality is a by-product of having a large brain, the behaviors arising from that knowledge (prayer, worship, confession, mourning) are not embedded in the gene. They are a consequence of experience, something we *learn*. As Frederic Prokosch writes in his novel, *The Asiatics*, "Death . . . is a specter discovered by man. Nature didn't intend us to know death so intimately." Indeed not all behavioral traits must necessarily be a product of natural selection, and this is ultimately a benefit to us. People can respond to environmental stimulation and circumstance in a range of ways.

A fair way of thinking about human behavior, then, is through the practical lens of adaptability. Just as we would not say that personality is a set of traits and all traits have corresponding genes, neither would we say that no features are tied to genes. Kite flying is not instinctual. Human sexual response ultimately is. The range is based on a preset genetic order, an order that functions in precisely controlled ways but that very much depends on the environment in which it is expressed. Most important, each person has a set of genes that is similar to but different (by as many as six million base pairs of DNA) from that of every other human creature, except a monozygotic twin, to whom he is identical. Each brings to the world a particular endowment, and each responds to the environment he selects in his own way.

Individual disposition is, once again, a flexible order, and denial of either side is limiting. If we have trouble seeing this for ourselves on a daily basis, we might do worse than to keep our eyes fixed on the evolving patterns of each individual.

Chapter Eleven

Nature, Nurture, and Implications for Psychotherapy

Know thyself.

Inscription on the Delphic Oracle

T he controversy over nature versus nurture has a long past. One could say psychology has a long past too, but only, as Ebbinghaus said, "a short history." Until this century, in fact, most descriptions of the mind were philosophical in character, awaiting the hard observations, empirical research, and direct biochemical and genetic study being done today. These are the data that make the short history of psychology, psychiatry, and psychotherapy, and a review of this relatively recent history reveals a compression of theories and insights that are often in competition with each other. What the argument so often involved, interestingly enough, was nature and nurture.

Just as teachers and parents need to understand individual behavior, psychotherapists need to know the extent of individual flexibility. They need to be able to recognize the biological and environmental forces that influence stability and change. Thus, the nature/nurture controversy is at the heart of the history of psychotherapy. The two histories shadow each other; often they intersect. The tension that has existed historically between heredity and environment has acted like a wedge in the field of psychotherapy, driving to one side or the other many of its theories and the techniques arising from them. In essence, in spite of Freud's effort to include *biology as well as experience* in explaining behavior and illness, we often consider only the power of experience. Now may be the best time to think anew about where we have been and where we are going.

Sigmund Freud's legacy, fifty years after his death, is alive with discussion about the influence of psychoanalytic thought and the meaning of his theories, even without the naturists or nurturists trying to lay claim to it. Yet interestingly enough, his struggle to understand the relationship of biology and experience may account for much of the controversy around him, and ultimately for his enduring influence.

Freud was trained in the medicine, brain physiology, and neuroanatomy of the nineteenth century. Many of his professors and contemporaries were looking for specific brain lesions, highly localized, as the key to behavioral abnormalities, and for the first time such mental illnesses as dementia praecox (schizophrenia), melancholia (depression), and hysteria were described. The very reductionism of this approach, a particular illness linked with a particular localized lesion, Freud ultimately found constraining – his observations of certain early patients did not confirm it – and so he loosened biological ties to pathological behavior enough to include great networks of experience. He was not the first person to do this. But he was one of the first with medical training who maintained the importance of constitutional factors in early childhood experience.

Watching Jean-Martin Charcot's demonstrations at the Salpêtrière Clinic in Paris, Freud for a time sympathized with the strictly physiological explanation Charcot advanced for hypnosis as well as the neurological explanation proposed for hysteria. At the same time another French scientist, Hippolyte Bernheim, introduced the notion that hypnosis was nothing more than a state of suggestibility, a purely *psychological* rather than a physiological phenomenon. Intrigued, Freud resisted either explanation at the expense of the other. He wanted both. "It would be just as one-sided to consider only the psychological side of the process as to attribute the whole responsibility for the phenomena of hypnosis to the vascular innervation [physiology]"; And "there are both psychical and physiological phenomena in hypnotism, and hypnosis itself can be brought about in one manner or the other."[1]

What remains Freud's legacy is a theory that links both experience and biology. On the one hand it describes the intricate network of influences in early childhood experience, and on the other it reveals the underlying forces of the drives that represent the biological demands on psychic life. This dual legacy has caused frequent misinterpretation, and what remains to permeate modern thinking are rarely both sides of it together.

There have been those, for instance, who misunderstood Freud's explanation of childhood sexuality and believed that a sex drive in children, biologically rooted, was both impossible and scandalous. There are others on the experiential side who believed that motherhood was the resounding refrain of his legacy, that mothers are the all-important influences in life, and that all who we are is tied one way or another to our mothers. And then there are still others who believed that Freud's central focus was on trauma, and that some external traumatic experience has to happen early in life for pathology to develop later on.

Misunderstandings of any theory are inevitable, of course, and perhaps some of the present confusion can be traced to the fact that Freud evolved and revised his theories right up to the time of his death. The Freud of 1899 is not the same as the one of 1922, or again of 1937, and to study Freud necessitates following the historical transitions in his thinking rather than isolating portions of it. In any case, the perspective we presumably enjoy today is one advantage his contemporaries could not have enjoyed when his theories were originally disseminated. For example, when Freud began to distance himself in the 1890s from purely biological explanations for hypnosis (and also for neurosis, as he broke with Josef Breuer), some associates perhaps seized too eagerly on his introduction of the idea of environmental influences on the psyche. Feeling suddenly freed from allegiance to theories of biological constraints, they embraced a theory they believed he meant to be entirely nonbiological. Frank Sulloway writes:

> From the very outset, the fashionable mystique that grew up around analytic psychotherapy derived largely

from the accidental, experiential considerations that were largely communicated about Freudian childhood. Patients and analysts alike were captivated by the environmental component of Freud's theories. Gone were the days of the hopeless diagnoses that had been so natural under the theory of hereditary degeneration. With Freudian psychoanalysis, a totally new era in the treatment of mental diseases had apparently arrived.[2]

Psychoanalytic theory, grounded in the exploration of early childhood experience, thus often created the appearance that it was a purely environmental explanation for the roots of personality and neurosis. Since the psychotherapeutic process itself revealed the repetition of these earliest patterns of behavior and the influences of early phase organization, complete with the reemergence of unconscious emotions and fantasies, this process presented the patient with a chance of resolving these old patterns and finding new modes of adapting to the world. His symptoms could be alleviated by working through the conflicts brought on by living. The patient's inner reaction to outer events was quickly considered pivotal to his therapy, whereas constitutional predisposition (the old biology), although recognized as a cause of some illnesses, was not in itself considered open to direct experience and so was less important (or unimportant) in psychoanalytic therapy. Psychoanalysis, it was believed early on by many of Freud's followers, was nothing if not a prescription for reshaping one's whole life.

Alfred Adler, one of Freud's early disciples who branched off to create his own school, was influenced by the then-prevalent idea that man's function is the result mostly of social forces. He wrote,

Investigators who believe the characteristics of an adult are noticeable in his infancy are not far wrong: this accounts for the fact that character is often considered hereditary. But the concept that character and personality are inherited from one's parents is universally harmful because it hinders the educator in his task and cramps his confidence. The real reason for assuming that character is inherited lies elsewhere. The evasion enables anyone who has the task of education to escape his responsibility by the simple gesture of blaming heredity for the pupil's failure."[3]

The corollary Adler hints at in this statement is that this evasion enables anyone who has the task of psychotherapy to escape his responsibility by simply blaming heredity for the patient's illness. This argument is a classic attack on biology's influence, claiming as it were that learning and change should be seen as limitless.

It is true enough that throughout history constitution has been blamed for many ills, but the damage caused by the misuse of biology cannot be remedied by disregarding its role in the human condition. Furthermore, although Adler's criticism of those who generalize too freely about heredity is right in one sense, the reason is not, as he says, because such a conclusion hinders the educator in his task. The endowed component of any individual at birth develops and adapts over time. There is nothing static in this process; it is free-moving and interactive, and educators will always have work to do, trying to stimulate all areas and recognizing the individuality of each child. This is equally true for the psychotherapist. We should not assume that a person's predispositions are the dimensions in life that inherently suppress or limit his living. They are often the carrier of the most advantageous forces. We may recognize constitutional disorder or illness as realistically limiting, but at the same time we must be willing as parents and educators – and psychotherapy must also be willing – to give support to the healthy individual's endowed inclinations, the set of talents and creativities that are uniquely his own.

In a letter to Carl Jung written as early as 1908, Freud mentions a colleague (not Adler) who he says "is undoubtedly a very interesting and worthy man, but . . . he is a fanatic. . . . He denies all heredity."

Adler was not alone, however. There were other important psychoanalysts who branched off from Freud and emphasized culture as the major determinant of the individual. In the so-called cultural school, Erich Fromm and Karen Horney, among others, separated themselves from the "biology" of Freud, from his theory of drives, and even from his emphasis on early childhood experience. As Stephen Frosh summarizes it, the "culture school theory, because of its focus on the way in which cultural conditions distort character . . . fails to provide an account of the manner in which society penetrates individual consciousness *in the process of its construction*" (italics in original).[4] In this way the individual is seen as passive, shaped by the social fabric and not shaping it in turn, not accommodating to it, as Piaget might have said. Culture alone cannot explain the intricate development of character and personality that Freud emphasized. He knew that a person's endowment reacts and adjusts to existing social conditions in its own way, influencing them in turn.

On a similar but not identical path as the members of the cultural school are the object relations theorists, whose influence on psychoanalysis in the last few decades has been truly significant. Since their findings have been based largely on wide-ranging studies of early infant development, it is not surprising that their main theoretic focus is the interaction between mother and child. They describe in wonderful detail the early capabilities of the baby, how it responds to the mother through gazing, early eye-to-eye contact, emotional cuing, and physical stimulation. Because the mother-child relationship is regarded as a unit by the object relations school, every nuance of development is seen in the context of that interplay.

Margaret Mahler's outline of the first three years of life, John Bowlby's attachment theories, and D. W. Winnicott's description of transitional objects have all been important contributions to the understanding of early development. It may be enough to say here that these formulations emphasized the child's interaction with and dependency on the parent during development and maturation. Thus, the biological influence of the drives and the constitutional contribution to personality development were deemphasized. "The emphasis of these theorists on the environment is thus not accidental: for them, there is no individual without the social, no self without the other."[5] The object relations school has sparked the acceptance of a number of propositions that have markedly influenced psychoanalytic technique, and it has contributed to an understanding of some of the earliest experiences of the child as they affect social development. But in addition to reducing the influence of biology on the child, object relations theorists also narrow the range of social influence beyond the mother, who is credited in this scenario as the nearly exclusive originator of health or pathology in her child.

The tendency to credit or blame the mother, assigning to her an overriding role in the development of the child, has its own long history. Mothers have been blamed for causing nearly every illness including schizophrenia (the "double-binding schizophrenogenic mother") and autism (the "refrigerator mother"). Without minimizing her role in development or the role she might play in contributing to pathology, at the same time we cannot see the mother as the lone culprit in what is really an essential partnership with others, including the father, other adults, and the child's siblings. In addition, by focusing on the earliest mother-child years, the later stages of development are diminished, almost ignored.

The many researchers who understood psychoanalysis as primarily an investigation into environmental influence alone have left large footprints. Their work on the many intricate environmental influences in development (social, cultural, and parental – including the minutiae of mother-infant interactions) has shifted the discussion away from the biological base Freud always maintained. In looking at some of these researchers and theorists, we can see how their positions crystallized around the issue of nature versus nurture.

We might add that although the mode of therapy we have been referring to is a psychoanalytic form of psychotherapy and psychoanalysis itself, the method most directly sculptured by Freud and his disciples, there are many other forms of therapy, cognitive and behavioral among them. They too describe the way a person mentally or emotionally experiences life, or reflexively reacts to it, and thus in them also is the appearance of the possibility of limitless change, which anyone who has ever participated in psychotherapy knows is not the case, nor – and this is most important – is it really the point.

When we consider the role of genetic endowment in life, the way it influences individual maturation and development and resurfaces across generations, we confront the question of how far psychotherapy can reach to alter the human condition and its conflicts. If every person brings to the world his own set of faculties that codetermine the very way he reacts to it, don't these inborn variations imply certain boundaries to flexibility beyond which change is impossible?

In one of his most important, closely reasoned papers, "Analysis Terminable and Interminable," Freud addressed himself to the question of the limits of psychotherapy, and thus to the question of nature and nurture itself. He identified two related ways in which a patient's analysis might end: "The patient shall no longer be suffering from his symptoms . . . anxieties and his inhibitions. . . . So much that was unintelligible has been explained . . . that there is no need to fear a repetition of the pathological processes." And "The analyst has had such a far-reaching influence on the patient that no further change could be expected to take place in him if the analysis continued."[6]

It is important to note in these related outcomes that Freud discarded the notion of making his patients "normal" or necessarily "happy." He rather believed that by gaining insight into conflicts and alleviating symptoms, the patient could then proceed to be who he is, relatively undisturbed by internal or external conflict. Only in this essential way could disposition assert itself, revealing the patient's individual temperament. "One feels inclined to say," Freud wrote, "that the intention that man should be 'happy' is not included in the plan of 'creation' . . . our possibilities of happiness are already restricted by our constitution."[7]

In this statement there is plainly a sense of the extent of what therapy can accomplish. Since conflicts are unavoidable and even necessary in development, it is the therapist's goal to bring internal biological demands into harmony with external influences, to bring the changing conditions of maturation and development in line with environmental conditions, and to co-ordinate the surging expression of drives with the cultural and social milieu of the patient.

Freud held that those pathological states caused by environmental influences, by trauma for instance, yielded better to therapeutic intervention than those determined by biological disposition, which were unchangeable by analytic therapy. Such a formulation is reasonable; it is realistic, because it recognizes the boundaries of change that are defined by the patient's individual makeup. Yet to the degree that it divides internal and external influence, such a notion may contribute to the unfortunate polarization of nature and nurture. As we have seen, external forces mold, encourage, or interfere with the expression of endowment, and endowment responds selectively to the influences of the environment. Instead of saying, as did Freud, that ego alterations "are either congenital or acquired . . . of these the second sort will be the easier to treat," we should reemphasize that it is the nature of the interaction of constitutional and acquired factors in all areas of mental function that determines the outcome of therapy. But this interaction still rarely gains the recognition it deserves.

Freud's contribution to the nature/nurture debate is thus often depicted only in caricature, sometimes on the side of biology, most often on the side of early childhood experience as the foundation of adult conflict and personality. But Freud himself never abandoned his belief in a span of forces. Just two years before his death he wrote, "Each ego is endowed from the first with individual dispositions and trends [and] before the ego has come into existence the lines of development, trends and reactions that it will later exhibit are already laid down for it."[8] He reconfirmed this biological dimension even as he described the role of trauma or emphasized the pivotal role the ego plays in regulating and mediating life. With his division of psychic structure into three parts – id, ego, and superego – he encompassed the biological, the reality and adaptational, and the ethical and social dimensions. In this way, he saw psychic inner life reflecting parental, societal, and cultural influences, while also remaining forever rooted in the biological matrix.

These functions of the ego, "endowed from the first with individual dispositions and trends," include an array of apparati (the perceptual, cognitive, and sensorimotor faculties we have seen) plainly linked to the constitutional givens at birth and to the maturational stages of life. The ego also mediates between drives and reality and so forges our very adaptation to and selection of environmental forces. This is what Freud meant when he said that there are biological demands on psychic life. This is what Joseph Campbell meant, speaking about the common ideas of myth, when he writes that "the psyche is the inward experience of the human body."[9] Body and mind inform one another in development: Freud saw that they cannot be separated.

It is easy to understand why constitutional forces are still so often neglected. First of all, aside from obvious hereditary illnesses, they are hard to recognize. Direct genetic mapping has now opened the way for investigation into the full range of genetic influences, but fifty years ago we had only personal observation and twin and adoption studies, a number of which were inconclusive or faulty (notoriously one by Cyril Burt). Not all but many early studies proposed direct links between heredity and, say, intelligence or gender, forging numbers and percentages as definitive results, with implications about race, sex, or nationality. This approach, as we have seen again and again, defies the logic of life's complexity. It describes a group's performance on a test without examining the influences that shape the distinctness of each individual in development.

Racism aside, no one wants to focus on an area over which he feels he has little influence, such as heredity. In view of Freud's position as well as the evidence gathered by infant observers, it is surprising that present-day psychological papers make next to no reference to the role of endowment. Case histories rarely begin with a description of the disposition, inclinations, and susceptibilities of the patient with which he then encounters his life experience. Whether the topic is depression or divorce or adaptation or work inhibition, conflicts are often described only in terms of the patient's psychic dynamics; and little or no attention is given to biological influences and the limits of adaptability.

This neglect is even more surprising as one realizes that a biological demand on the psyche does not imply that there is one and only one path for each individual in life, a path that is determined by his genes. "Genes set limits for ranges," we have seen; "they do not provide blueprints for exact replicas." A biological demand on the psyche implies the existence of a predisposition to respond to the environment in a way unique to each person. The limits in this process are less constrictions of freedom than avenues for flexibility – not the flexibility to become someone wholly new, but rather the flexibility to explore the manifold branching of possibility that truly exists in nature and that life's experiences can mold. Psychotherapy assists in that exploration.

If there are limits in our lives, limits like concrete boundaries, they are as easily encountered in the environment as in our bodies: poor nourishment and clothing, missing parents, bad teachers, limited choice of jobs, an unloving mate. The contribution parents and professionals make when trying to understand development, support it, put it on track if it becomes derailed and make it flower, is plainly grand. Every effort must be made on behalf of those in need – not blindly, but rather in recognition of the individual in all the aspects that make him human, both biological and experiential.

"Know thyself" was one of the inscriptions on the Oracle of Delphi. As Stephen Jay Gould remarks, it is also the basis of the name Linnaeus gave to us as a species: *Homo sapiens*, Knowing Man. When we come to know who we really are, blend of nature and nurture, only then are we truly knowing.

PART IV

To glimpse one's own true nature
is a kind of homegoing . . .
the homegoing that needs no home.
PETER MATTHIESSEN, *THE SNOW LEOPARD*

CHAPTER TWELVE

Living without Grandparents: The Loss of Intergenerational Transmission

*Generations pass while some tree stands,
and old families last not three oaks.*
SIR THOMAS BROWNE

The structure of the family has changed. It has been reduced from an extended unit of many living branches and generations to a nuclear constellation of mother, father, and children – and reduced even further by divorce. In 1960, 9 percent of all children grew up in single-parent homes. By 1986 the number was up to 24 percent, and in families below the poverty line it was 50 percent. With fewer members in each family, the transmission of knowledge and tradition over the years becomes uncertain. Without an oral history, the patterns revealed by a family across generations disappear, and with them goes the knowledge of the characteristic ways in which past members responded to life's situations. The living experience of the differing paths their lives followed, unfolding gradually, now becomes if not unknown, then known only by hearsay.

Grandparents once had a pivotal role in this process, a role that went far beyond the purely physical support they gave their grown sons and daughters to help ease the pressures of parenthood. With knowledge of their own parents and grandparents plus their children and grandchildren, they had the perspective of at least five generations. They provided an oral family history of achievements, illnesses, and failings; they knew in what ways a new grandchild might or might not fit previous family patterns; they knew other family members who were slow in some areas of development, or had short tempers, or developed successful ways of compensating for their weaknesses. Grandparents were thus the ones with the knowledge of what helped or hindered life's development, and they were the ones with the experience to know how sensitive a person should be toward it. In sum, they knew the family's past literally from inside and out, both the hereditary history and all the reactions to the experiences of living.

Not all grandparents are everything we describe, of course. Perhaps much family information is simply unavailable to them, or their impressions are tainted with subjectivity. Given the fact that few families today have grandparents living in the same state, much less the same neighborhood or home, the contact necessary for even a limited recounting of lives from person to person is diminished even more. Few families can afford homes large enough to support extended members; fewer still would want to if they could. This is the reality. Yet we also wonder what we miss because of it. As living witnesses to five rungs of the family ladder, grandparents – or even better, great-grandparents – have more information than anyone ever taps. Their relationship with each member is unique, and if they could be encouraged to keep a family diary full of their recollections of all the relations on all the branches that they know, their effort in reviewing and recalling this data might trigger an even greater, more intricate variety of memory than anyone expects. Start, for instance, with obvious information about inherited disease. Just as some countries, especially in Scandinavia, have begun compiling family data on medical computers, a grandparent's diary could help trace the course of family traits – not only what occurs but also when it occurs in the life of each family member. With such a tool the appearance, reappearance, or absence of cancer, stroke, and heart disease, or the flux of alcoholism, dyslexia, and anorexia could be followed. Such a search could potentially reveal startling data, of the kind only professional researchers could match. Extreme genetic conditions can actually be traced to their hereditary origins, as Jared M. Diamond illustrates in "Founding Fathers and Mothers":

> Why, for example, are there 82 six-fingered dwarfs among the Amish population of Lancaster County, Pennsylvania? Because the few founders of that community included a certain Samuel King, and either Mr.

King or his wife happened to have the gene for six-fingered dwarfism. Why do many South Africans, but few people elsewhere in the world, have a genetic condition called *osteodental dysplasia* that causes all their teeth to fall out by the age of twenty? Because all those South Africans are descended from a polygamous immigrant named Arnold, who carried the gene and spread it among the 356 traceable descendants he sired by his seven wives. Why is Huntington's chorea, a fatal neurological disease, disproportionately common in Tasmania and southwest Australia? Because an English widow called Miss Cundick, who emigrated to Australia in 1848 with her thirteen children by two marriages, passed the gene to offspring of both marriages and thereby became the ancestor of at least 432 Australian victims of Huntington's chorea.[1]

If genetic diseases can be traced systematically through an oral history, why not a host of other family characteristics also, including the faculties of perception, cognition, sensorimotor skills – and ultimately personality? This sort of information should be as important to parents, educators, and therapists as a major art retrospective is to the art critic. Without such a context, an individual at a given point in his life, like an individual painting, is difficult to judge. Yet the benefits we accrue by tapping this information also lie in less tangible areas. In listening to grandparents, children are often attentive and curious because they have a window on the lives of those around them. They hear what life was like for the members of their family, how relatives acted in response to events or were overwhelmed by them, how brothers, sisters, parents, aunts, uncles, and ancestors behaved; and through it all they are given the chance to identify with – or differentiate themselves from – this assortment of relatives.

This process is crucial for the child, for it marks a step toward individuation. And so we reach the deeper value of the extended family – and the great efforts that must be made in its absence.

Through individuation the developing child moves toward establishing a sense of self. Comparing himself to those around whom he lives is especially important in this process. Imitating others and internalizing their experiences – wanting to be or do like them – establishes bonds that are an integral part of belonging. Once these bonds of likeness are secured, separation, individuation, and an acceptance of others may become possible, leading to an even higher level of bonding.

Perhaps we make it sound easier than it is. The developing road to selfhood through individuation is actually full of rivalry, envy, and jealousy; it may falter from lack of care or relentless overstimulation; and there may be an absence of suitable, fitting role models. Fortunately, however, an extended family confers benefits on the child for reasons that involve both parts of the equation, *extended* and *family*.

First, only in an extended configuration will children have a choice of relationships and a variety of temperaments with which to interact. In a nuclear family or a single-parent home (often just mother and child) there are obviously fewer possible relationships, and those that exist may be too severe, inescapable, or not nurturing enough either by neglect or over-intensity.

But the child in a larger family can escape mother and father's angry mood or undue expectation by allying himself with other family members who may be more attuned to his needs – often a grandparent, but also an uncle, a cousin, and so forth. The new pair may suit each other, understand each other, and use each other better. The selection is not always of those temperaments that are alike, or those who share the same preferences or weaknesses; the weaker may also attach to the stronger, the isolated to the outgoing, the entertainer to the thinker, and so on. The combinations are endless, even though the attachments may be only transitional. The favorite grandparent, aunt, or sibling may be the one who most closely matches the child's needs at a given time in development, someone who is cuddly when the child is cuddly, independent when he is independent. However, needs also change, and although a child may desire close matching to mother in infancy, a matching so close it approximates symbiosis, the opposite may be true in adolescence: he may seek and idealize the person who seems least like himself, someone strong because he feels weak, good because he feels bad, bad because he feels too compliant; a person, in short, so much unlike his parents that he can detach himself from their influence. To restate Hartmann, adaptation is based on each person's selection of the environment that suits him.

People often remember how important these relationships were to them when they were young – the rapport, the confederation, even the intimacy that may last over years until new developmental conditions favor a change.

From infancy onward we can observe the choices people make, choices that say much about their dispositions. We can't expect all members of a family, developing together, to be alike. Early faculties differ in people, and the differences evoke varying responses. A baby girl may not be able to be soothed by her mother, yet when a grandparent or aunt holds her she will stop crying. "There are indications that mothers are more comforting, whereas fathers lean more toward activation and stimulation of their children. [However,] when both parents are available, the children may turn to one or the other depending on their needs at a given time or on their disposition and stage of development."[2]

Other members may be even better attuned to such a child, at least at that time, knowing how tightly to hold her, at which pace to rock, what pitch of voice to use, and so on. At age two or three the normal negativism that the child holds toward a parent – rejecting food or resisting toilet training – may be less intense or absent in her relationship to an aunt, for whom she may be willing to eat and find the toilet by herself. This flexibility partly accounts for life's variety. Needs change, and perceptions change with them. What we encounter in the world as we develop adds layer upon layer of new meaning for us.

Since our individual endowments incline each of us uniquely to respond to the variety of experiences we encounter, in an extended family a member may more easily recognize the human variations of which he is one. Children in such families may be in a better position to differentiate themselves, and grandparents may offer the wisdom of both perspective and support. The burden of understanding, of course, finally falls on the parents, who are responsible for the child's overall care. It is to parents that the observation of individual relationships and variations can be most helpful, guiding them in an understanding of their family and the modes of helping that may be most appropriate and best timed. But knowingly or unknowingly, family members such as grandparents bring with them a perspective that initiates such insights. This often endless reevaluation opens children up to a needed comparison with others, contributing to the evolving profile of each new child.

There is one additional benefit of extended family connections – to the grandparents themselves. Not only can they feel needed – still or again – but they may take advantage of their relationships in the same way children do, learning about and checking themselves against the variety of lives around them. They can rediscover themselves, their strengths and weaknesses, and give an example to their grandchildren not only of their collected knowledge and independence but also of the benefits of caring. Their presence is a model not of isolation but of interdependence, and it strengthens the ties between all family members.

We have been dreaming, of course. So few families today, whether because of social changes or economic hardship or both, can afford the home and food to keep extended family members together. However, in the absence of an extended family, parents should be aware that they are presented with an ever broader challenge – *to become more flexible.* Parents must learn to see each child in his own pattern of developmental unfolding, and they can learn to offer a supple, responsive environment to the changes bound to occur in each child over time. They must be flexible enough to provide the varied support other family members would have provided had they been present. And yet the task is often more difficult. Now that more mothers go back to work soon after their children are born, it is often up to the nurse, the housekeeper, the group worker, or the nursery school teacher to decipher the particular needs of the individual child. What we do not allow grandparents to do anymore, we assign to outsiders.

Grandparents, for their part, have often learned their lesson too well. They often accept their children's reluctance to make them a part of the family life, becoming by necessity more distant, moving to better climates if they can afford it, ultimately dwelling in their own freedom.

When we view this scenario from the perspective of intergenerational disruption, a disruption of experience and accumulated knowledge, of an understanding of family biology, family identity, and family dynamics, the ensuing loss seems irreparable. Perhaps it is not. Parents alone, however, must work especially hard to accept and foster the wide expression of their children's developing individuality, an individuality born of nature and nurture.

We have looked at the advantages conferred on children growing up in extended families – with emphasis on the word

extended. Now a note on *family.* One of the questions that is
often asked is how a child develops if the caregivers around
him are not family members. Just how narrow or wide is the
child's range of response to other people? Will one person do
as well as any other?

Sally Provence and Rose C. Lipton wrote in 1962: "The
family . . . is the setting in which babies can best be provided
with the care and influences that support and foster good development. It becomes increasingly hard to support such care
as we move farther away from this model. The infant's needs
are multiple and complex, and it is difficult and perhaps impossible to meet them adequately under conditions of group
care."[3] Although group and institutionalized care has become
generally better since the time that statement was made, it still
may not equal that of the family. As Provence added in 1989,
"Institutionalized care has rarely been so designed and provided
that it meets the physical, social, emotional, and cognitive needs
of infants and young children in a way that approaches the
nurturing, stimulating, and protective atmosphere of a good
family environment."

Families are better nurturers than anyone else for the
primary reason that they can provide better-tailored care, care
more appropriate to the individual child. An adoptive parent,
or an aunt or uncle who permanently takes over the role of
parent, will have similar advantages. But if we assign some of
the functions of rearing to group care, we meet new difficulties.

From time to time in history, the notion of a "community of man" has gained enough popularity to spawn groups
hoping to bring the idea into the social arena, the workplace,
the home. We speak here not only of groups grounded in social
and political ideology – socialists, for instance – but also of
smaller movements whose members hope to share their lives
and make responsibilities and benefits collective, including the
responsibility of child rearing.

Communes still exist today, some of a religious nature; yet many of these experiments simply never survived: think for example of the communes formed in the 1960s in the United States. The kibbutz system in Israel has lasted for several generations – but not without changes, and the changes are quite significant.

From the time kibbutzim were first founded in the late nineteenth century the children of kibbutz members lived in a group physically apart from their parents, though parents could visit after work and bring the children home for a few hours. The rationale for this separation was both practical and ideological. It freed the parents to work without worrying about their children's well-being, and in principle it fostered the growth of a "collective" person, one who defined himself in relation to the group. The values the children developed – "what's good for all is good for one" – were meant to be an improvement over those of capitalist, competitive societies.

Many of the children who grew up in these collectives, however, are today pressing the kibbutz society to shift the responsibility for child rearing back to parents. Now parents themselves, they want to take personal care of their children after work and have them sleep at home rather than with the group teacher in the group house. They want to be more like the traditional parents that the movement left behind. Perhaps one reason for this desire is the almost inevitable readjustment of ideological fervor on the part of the younger generations. With the pioneer spirit no longer needed to survive, they may feel less reason to give up their children at night. But another reason may lie deeper still.

It was found that children in kibbutzim who lived and matured among their peers still primarily organized their early inner lives in reference to their parents rather than to the group. As adults, they now speak about the significance of their mothers and fathers – of being preferred by them or neglected, of their love and resentments, and of their rivalry with their natural brothers and sisters rather than with the other children in the group. Even the significance of the group teacher, the *metapelot*, is overshadowed by the memory of their parents.

It seems that we have an inner striving or hunger to establish a psychological belonging to our true families, a belonging that cannot be suppressed by any ideological experiment. Ideology may take us away from ourselves for a time, as we have said, but it is as if some self-regulatory natural guide brings us back and reasserts itself. Perhaps the climate that has relegated the extended family to history and created the single-parent child will one day, by choice or necessity, begin to reverse itself. A larger family unit, if not wholly reinstated, may at least be longed for, not only because of the need for social and economic support but also for compelling psychological reasons.

The reduction in family size to the smallest of units may limit, or even eliminate, the knowledge of intergenerational family history. In its absence we can create new "families" from among friends and strangers, but the anchor that a real family provides for succeeding generations and the place each member has in its continuum is not easily replaced by liaisons with outside friends, even good and intimate friends. For the child, although there may be a tendency to move away from the family to other relationships when conflicts arise, the adaptation to and tolerance of differences between himself and his family is a meaningful step in development. The acceptance of family fosters acceptance of everyone else. If differences are not worked through in this context, can we expect them to be resolved as easily later, outside the family?

With or without the help of grandparents and others, it is parents' most basic obligation to be flexible enough to recognize the child in his individuality – to learn to give him what he needs in order for him to become, with every flourishing variation, who he is.

CHAPTER THIRTEEN

Individuality and Groups: A Final Consideration of the Human Reliance on Others

Life does not make any sense without interdependence.

ERIK H. ERIKSON

E rikson's dictum turns on the recognition of the individual both alone and within a web of relationships. Yet
as human beings we often have difficulty understanding
our lives both separately from those around us as well as in
coordination with them. Given the marked variety of human
life, the myriad cultures shaped by history and geography, the
codes by which we have chosen to live are many and imper-
manent. There has never been a *right* answer for mankind, at
least never a single answer. We have been nomads, clinging to
small numbers: we have been solitary searchers, priests, and
ascetics, wandering alone or with God alone; we have been
hunter-gatherers, farmers, and villagers; we have been mass
city dwellers, living among millions of others yet never more
distant from our neighbors.

One explanation for this variety lies in our amazing
flexibility as a species. Evolution has made us adaptable; we
can survive and reproduce in an extraordinarily wide range of
conditions. Not in an infinite variety – not, perhaps, in the
darkness of a nuclear winter. But perhaps even in that, too.
Adaptability has been built into our constitutions, and the very
genes that program our individual dispositions also bring a
demand for the society of others, for their nourishment, their
stimulation, and their ability to help us gage and fulfill our
understanding of ourselves. Whereas other animals depend on
their parents for a few days or weeks or months, we depend
on ours for years, both physically and emotionally. In some
ways many of us never outgrow that need.

Some biologists interpret all life through the gene, that
is, through the biochemistry of life as directed by the genetic
matrix. This biochemistry alerts glands to secrete hormones
that enrage us, sadden us, or sexually excite us. It is a potent
force, indeed. The particular combination of genes that com-
poses our individual genotype makes us at birth different from
anyone else (except an identical twin). It distinguishes our de-

velopment and maturation throughout life, punctuated by the changes our genes help to produce, such as learning to walk or read, the beginning of puberty, then aging and death. The development of our personalities is also connected, though not shackled, to this genotype. The title of Robertson Davies's novel *What's Bred in the Bone* is taken from a Latin proverb that holds some truth: "What's bred in the bone will not out from the flesh."

Our inborn need for others, however, as well as our adaptability to the conditions we meet, takes us away from a purely genetic view and toward one of interrelationship. An individual must be able not only to assert his own needs but, to a greater or lesser degree, to orchestrate his needs with those of others. The presence of strong and influential groups raises some most interesting questions in this regard.

By *group* we mean any number of people who exert pressure and bind the individual to them for any of a variety of reasons. A group can be dictated by political ideology, run by states or entrepreneurs, or formed by social need or religious determination, and for days or decades it may command individuals to serve it. Masses of people may sacrifice their individual needs and submerge themselves in the group, accepting unquestioningly its ideals – those of Nazism, for example, or of Hari Krishna cults, or of social and political revolutionaries. Under the auspices of a transcendent ideology people can be made, or allow themselves to be made, to do things that under other circumstances, at other times, they would regard as abhorrent, or perhaps only useless. They yield their own predispositions, their own needs and rhythms, and let the surge of group fervor sweep over them, for an hour or a lifetime.

There is a large literature on mass psychology. We wish to stress here only that extreme social pressures – as common today as they were in the past – exact a price on individuality. Under the charismatic force of a leader we may suspend our personal patterns of living and do as the group does. When the leader dies or the movement dissolves, members begin again to be themselves, and the wave of confusion about what to do without the group's influence can be replaced by a sheer sense of terror – or by a wave of relief and a new sense of who one is.

Interestingly, we may have a built-in tendency to submerge ourselves in groups, like children who rush under the parental umbrella. It may serve us well, perhaps even help us survive. But some of us seem to conform to group movements much more readily than others. While there are bureaucrats whose job it is to attend to society's rules, there are also loners and artists among us. Individuals span a wide range in their ability to adapt. Some of us stay on a plane of dependency and submersion; others follow a path toward individuation and the fulfillment of nature's maturational and developmental plans. Some people return quickly to themselves after the experience of merging, whereas others remain forever shaped by the collective mold.

We don't yet know enough about groups, and by extension society and culture, to understand what happens to individual dispositions under their influence. We can see that genetic influences on individual development can be stifled by harsh group teaching. At the same time, endowment cannot do without teaching and interaction. Our dispositions are honed and revealed by the groups known as family and society. We are all behind in our understanding of how familial, social, cultural, political, and global forces move and shape each of our lives, not to mention how these forces can be tailored to fit our individually endowed patterns.

Group ideology can take us away from ourselves, at least for a time. The trends inherent in nature may bring us back, comfortably returning us to a freedom to be as we were; but solitude is not the state in which we feel completely ourselves. Just as total individual submersion in a group is not ultimately natural, neither is rigid independence. People need others to shape themselves. In addition, society and culture need the participation of their members. It is not surprising that when masses of people become overly independent, withdrawn, or self-absorbed, the tissue of society turns fragile. Civil wars have resulted when the multiple states of a union choose their own paths over the collective one, and at such times we yearn to hear the words that will bond us to one another again: "A house divided against itself cannot stand," or "Ask not what your country can do for you: ask what you can do for your country." Such words reattach us to the society of others, to which we are as much inclined by nature as we are to our own separateness and individuality.

When the interdependence of individual and group is recognized, by ourselves, our parents, our teachers, and our healers, an interdependence born of seed as well as soil, nature as well as nurture, the meaning of individual disposition itself is enhanced. We must find a balance between too much interference in the natural rhythms of our lives and too little connection with the outside world for those rhythms to resonate. Perhaps we can then begin to see ourselves as simultaneously biological and social beings, for in whatever direction development steers us we will not find conditions that split nature and nurture into opposing forces.

The process will begin slowly. We again quote Tennyson's verse: "The baby new to earth and sky, / . . . Has never thought that "this is I": / But as he grows he gathers much, / And learns the use of "I" and "me," / . . . So rounds he to a separate mind / . . . His isolation grows defined."

This isolation is not inviolable. The recognition of the true sense of "I" – as bearer of inherited tendencies and receiver of the nourishment and guidance that only others can bring – allows for the "I" to reach out, not only in need, but also in compassion and sympathy. Leonard Woolf writes in *The Journey Not the Arrival Matters*:

> It is only if you feel that every he or she has an "I" like your own "I," only if everyone is to you an individual, that you can feel as Montaigne did about cruelty. It is the acute consciousness of my own individuality which makes me realize that I am I, and what pain, persecution, and death means for this "I."

Taken a step further, we may reach for a moment – but what a moment – a state of true empathy with the world, able to see it at once as a unity – and as a unity comprised of individuals. It is this state that Dickens dreamed of, against the selfishness and greed of modern society, in the hope that we can "think of people . . . as if they really were fellow-passengers to the grave, and not another race of creatures bound on other journeys."

We are, each of us, separately determined and uniquely configured. Yet our ability to adapt means that we also present ourselves to the world – family, society, and culture – to be shaped, guided, and renewed for as long as our development requires it of us. We must not say of a person, "What's bred in the bone will not out from the flesh," but rather, in recognition of all the work that life's natural program demands from him, from birth through development, echoing Goethe: "That which he has inherited he has made his own."

Epilogue

Old Hollywood B movies about French revolutionaries, benign pirates, and Robin Hood's thieves usually have a scene around a fire at night, when the eager-faced leader (à la Errol Flynn) asks a new recruit: "Well, which will it be . . . are you for us – or against us?" The question draws hushed silence from the campfire crowd. It cuts right through to the bone of allegiance, forces a decision on the listener, demands a stand, and ignores unnecessary nuances.

Readers of this book may well want to ask a version of this very question now: "After all is said and done (so don't mince words) . . . on which side do you finally stand – nature or nurture?" adding wisely, "And don't simply say interaction. That's too vague."

"Interaction," when used almost casually to avoid answering the first question, *is* too vague, of course. It is too simple: but it is not wrong.

We have tried to show that if one wants to expose the roots of individual behavior and temperament, individual maturation and development, individual health and illness, and individual adaptation to the world, one has to look past the lumbering categories of nature and nurture and into the eye of their interrelationship – which is unique for each person at each stage of his life.

What we have tried to do in these pages is to redraw the map of the nature/nurture controversy to emphasize its extraordinary complexity, since one-word answers such as heredity or environment – or interaction – no longer suffice.

The various case studies cited in this book illustrate the nuances of that complexity: how some inherited susceptibilities (for certain diseases, for instance) may pierce through any environment; how other tendencies rely on specific outside encouragement in order to flourish – and appear in some people but not in others; and how certain environmental situations can overpower any person's natural resistances.

Still, the reader may wonder: "Doesn't this explanation oscillate too freely from one side to the other? What if my daughter has trouble in school . . . should I look to her teacher as the cause, or to her genes, or to my ability as a parent? There must be specific answers to these questions."

There are answers, often good answers, but again they are complex rather than simple. And we can be sure, based on all we know today, that inherited tendencies play a part.

When one observes children in clinics from earliest infancy through maturity; when one does therapy and sees the way each individual brings specific predispositions and sensitivities that influence the way he adapts to the world; when one considers the timetables set by genes for so many physical and emotional milestones, from learning to walk or read to the onset of puberty and aging; and finally, most startlingly, when one confronts the powerful breadth of new DNA research, genetic recombination and chromosome-mapping of both mental and physical illnesses (and soon normal traits as well), one will not be satisfied with the purely environmental explanations for life that are too often given.

Despite all we have learned about the role of genetic makeup in development, people (health professionals, teachers, writers, and parents among them) have preferred the limitless potential that is offered by learning over the apparent constraint imposed by genes. Still reeling from the blows inflicted by such forces as Nazism, social Darwinism, and racism, a truly fair society such as the one in which we hope to live has become wary of hereditary-based arguments that echo of injustice.

It is time, however, to reconsider the actual role of

heredity in our lives: rarely as a fixed determinant of life (for genes also code for flexibility and adaptability), not as an accessory to genocide (genetic knowledge can be misused, but genes themselves are DNA codes and have nothing to do with ideology and politics); not as a limiter of potential (genes may allow for the most fabulous combinations of talents and temperaments, creating in each of us so many new possibilities); and not as a force opposing environment, for interaction with the environment occurs from the beginning of life through the moment of death. Although we want to combat the racism that emerges from the misuse of heredity, we must finally come to terms with what we now know: our inherited susceptibilities and tendencies help determine who we are as individuals; they set the very ground plan for our maturation and development.

In order to achieve personal insight, we must each become aware of heredity's role in our lives. We must each discover our inner ranges, potentials, and tendencies. It should simply become one of the goals of growing up.

The questions we ask in the morning about nature and nurture are often answered by afternoon. Follow the newspaper and keep up with the complexities of life as they slowly unravel: the causes of diseases and obsessional behaviors, the origin of emotions and personality. We are at the threshold of new ways of thinking about ourselves that can no longer be ignored.

Though we may settle each evening with an aggregate of influences, a collection of weblike strands that leave complex trails to both our genes and our experiences, future unmarked trails will lead to combinations of the two at which we can now only guess. Examine each life for its surface meaning and run into these strands by the thousand. Look backward and see a childhood of parental influence – but still further back to find generations of inherited connections. We are bound to the strands that lead us back, as we are bound to tradition.

But what is a tradition if not a slow spiral of change? Though our biologies and cultures repeat the past, they are open to change through adaptation. If each of us is literally a new experiment of nature, we can do no better than to recognize who we are in order to best plan where we might go.

Notes

PART I

CHAPTER TWO
The "Stranger" in Our Midst

1 D. W. Winnicott, *The Child, the Family, and the Outside World: Studies in Developmental Relations* (Reading, Mass.: Addison-Wesley, 1987).

2 T. Berry Brazelton, *Neonatal Behavioral Assessment Scale,* 2nd ed. (Philadelphia: Lippincott, 1984).

3 Erik H. Erikson, *Childhood and Society* (1950; New York: Norton, 1963), 69.

4 D. W. Winnicott, *Through Paediatrics to Psychoanalysis* (London: Hogarth, 1958), 163.

CHAPTER THREE
Traits and Personality: Carving Nature at the Joints

1 A. Tellegen, D. T. Lykken, T. J. Bouchard, K. J. Wilcox, N. L. Segal, S. Rich, "Personality Similarity in Twins Reared Apart and Together," *Journal of Personality and Social Psychology,* 54 (1988): 1031–1039.

2 Gordon W. Allport, "Traits Revisited" (1966), in *Readings in Personality,* ed. Harriet N. Mischel and Walter Mischel (New York: Holt, Rinehart and Winston, 1973), 21.

3 Walter Mischel, "Continuity and Change in Personality," in *Readings in Personality,* ed. Mischel and Mischel, 83.

4 Allport, "Traits Revisited," 20.

5 C. G. Jung, "Psychological Types" (1924), in *Readings in Personality*, ed. Mischel and Mischel, 24.

6 Jonathan Green, Martin Bax, and Helen Tsitsikas, "Neonatal Behavior and Early Temperament: A Longitudinal Study of the First Six Months of Life," *American Journal of Orthopsychiatry* 59, (January 1989): 82–93.

7 Leslie Brothers, "A Biological Perspective on Empathy," *American Journal of Psychiatry* 146 (January 1989): 17.

8 Donald J. Cohen, et al., "Personality Development in Twins: Competence in the Newborn and Preschool Periods," *Journal of the American Academy of Child Psychiatry* 11: 4 (1972) 625–44.

CHAPTER FOUR
Timetables of Change: Maturation

1 Susan L. Farber, *Identical Twins Reared Apart: A Reanalysis* (New York: Basic Books, 1981), 243–44.

2 A. Gesell and H. Thompson, quoted in Theodosius Dobzhansky, *Mankind Evolving: The Evolution of the Human Species* (1962; New Haven: Yale University Press, 1975), 61.

3 Margaret Mahler, "Rapprochement Subphase of the Separation-Individuation Process" (1972), in *The Process of Child Development*, ed. Peter B. Neubauer (New York: New American Library, 1976), 215–30.

CHAPTER FIVE
Timetables of Change: Development

1 Erik H. Erikson, *Identity: Youth and Crisis* (New York: Norton, 1968), 92.

2 Theodosius Dobzhansky, *Mankind Evolving: The Evolution of the Human Species* (1962; New Haven: Yale University Press, 1975), 60.

3 Jean Piaget, quoted in Richard F. Kitchener, *Piaget's Theory of Knowledge: Genetic Epistemology and Scientific Reason* (New Haven: Yale University Press, 1986), 40.

4 Erikson, *Identity: Youth and Crisis*, 93.

5 Dobzhansky, *Mankind Evolving*, 60.

6 Serge Lebovici and René Diatkine, "Normality as a Concept of Limited Usefulness in the Assessment of Psychiatric Risk," in *The Child in His Family: Children at Psychiatric Risk*, ed. E. James Anthony and Cyrille Koupernik (New York: Wiley, 1974), 11–12.

7 Howard P. Rome, "Personal Reflections: Personality Disorders," *Psychiatric Annals* 19 (March 1989): 121.

8 Susan L. Farber, *Identical Twins Reared Apart: A Reanalysis*, (New York: Basic Books, 1981), 235–36.

PART II

CHAPTER SIX
The Bridge Called Adaptation

1 Heinz Hartmann, *Ego Psychology and the Problem of*

Adaptation (New York: International Universities Press, 1958), 26.

2 Jean Piaget, quoted in Richard F. Kitchener, *Piaget's Theory of Knowledge: Genetic Epistemology and Scientific Reason* (New Haven: Yale University Press, 1986), 112.

3 Anna Freud, *The Ego and the Mechanisms of Defense,* rev. ed. (New York: International Universities Press, 1966), 120.

4 Hartmann, *Ego Psychology,* 23.

5 Anna Freud, *The Ego and the Mechanisms of Defense* (1937; New York: International Universities Press, 1966), 51.

CHAPTER SEVEN
The Vulnerable and the Invulnerable

1 E. James Anthony, "A New Scientific Region to Explore," in *The Child in His Family: Vulnerable Children,* ed. E. James Anthony, Cyrille Koupernik, and Colette Chiland (New York: Wiley, 1978), 5.

2 Donald J. Cohen, "Competence and Biology: Methodology in Studies of Infants, Twins, Psychosomatic Disease, and Psychosis," in *The Child in His Family: Children at Psychiatric Risk,* ed. E. James Anthony and Cyrille Koupernik (New York: Wiley, 1974), 387.

3 Albert J. Solnit, "The Vulnerable Child – In Retrospect," in *The Child in His Family: Vulnerable Children,* ed. Anthony, Koupernik, and Chiland, 645.

4 Emmy E. Werner, "The Children of the Garden Island," *Scientific American,* April 1989, 106–111.

Part III

Chapter Nine
A Long History Briefly Told

1 J. B. Watson, *Behaviorism* (New York: Norton, 1924), 11.

Chapter Ten
Flexibility in an Ordered World

1 Melvin Konner, *The Tangled Wing: Biological Constraints on the Human Spirit* (New York: Harper Colophon, 1982), 65.

2 Stephen Jay Gould, *Hen's Teeth and Horses' Toes* (New York: Norton, 1983), 169.

3 Stephen Jay Gould, "Of Kiwi Eggs and the Liberty Bell," *Natural History*, November 1985, 22–4.

4 Evelyn Fox Keller, *Reflections on Gender and Science* (New Haven: Yale University Press, 1985), 171.

5 Stephen Jay Gould, *The Mismeasure of Man* (New York: Norton, 1981), 333.

6 Edward O. Wilson, *On Human Nature* (New York: Bantam, 1979), 33.

Chapter Eleven
Nature, Nurture, and Implications for Psychotherapy

1 Sigmund Freud, quoted in Frank Sulloway, *Freud: Biologist of the Mind* (New York: Basic Books, 1979), 48, 50.

2 Sulloway, *Freud*, 438.

3 Alfred Adler, quoted in Dobzhansky, *Mankind Evolving: The Evolution of the Human Species* (1962; New Haven: Yale University Press, 1975), 14.

4 Stephen Frosh, *The Politics of Psychoanalysis* (New Haven: Yale University Press, 1987), 95.

5 Ibid., 108.

6 Sigmund Freud, "Analysis Terminable and Interminable" (1937), *Standard Edition* 23:8, 1964.

7 Sigmund Freud, quoted in Frosh, *Politics of Psychoanalysis*, 68.

8 Sigmund Freud, "Analysis Terminable and Interminable," *Standard Edition* 23:209, 1964.

9 Joseph Campbell, *The Power of Myth* (New York: Doubleday, 1988), 51.

PART IV

CHAPTER TWELVE
Living Without Grandparents: The Loss of Intergenerational Transmission

1 Jared M. Diamond, "Founding Fathers and Mothers," *Natural History Magazine,* June 1988.

2 Peter B. Neubauer, "Fathers as Single Parents," In *Fathers and Their Families,* ed. Stanley H. Cath, Alan Gurwitt, and Linda Gunsberg (Hillsdale, N.J.: Analytic Press, 1989), 68.

3 Sally Provence and Rose C. Lipton, *Infants in Institutions* (New York: International Universities Press, 1962), 1.

Selected Bibliography

Adler, Alfred. *Understanding Human Nature.* 1928. New York: Fawcett, 1957.

Ainslie, Ricardo C. *The Psychology of Twinship.* Lincoln: University of Nebraska Press, 1985.

Allport, Gordon W. "Traits Revisited." 1966. In *Readings in Personality,* ed. Harriet N. Mischel and Walter Mischel. New York: Holt, Rinehart and Winston, 1973.

Anthony, E. James, and Cyrille Koupernik, eds. *The Child in His Family: Children at Psychiatric Risk.* New York: Wiley, 1974.

Anthony, E. James, Cyrille Koupernik, and Colette Chiland, eds. *The Child in His Family: Vulnerable Children.* New York: Wiley, 1978.

Barash, David P. *Sociology and Behavior.* New York: Elsevier, 1977.

Bowlby, John. "Developmental Psychiatry Comes of Age." *American Journal of Psychiatry* 145 (January 1988): 1–10.

Brazelton, T. Berry. *Neonatal Behavioral Assessment Scale.* 2nd ed. Philadelphia: Lippincott, 1984.

Brothers, Leslie. "A Biological Perspective on Empathy." *American Journal of Psychiatry* 146 (January 1989): 10–19.

Campbell, Joseph. *The Power of Myth*. New York: Doubleday, 1988.

Cohen, Donald J. "Competence and Biology: Methodology in Studies of Infants, Twins, Psychosomatic Disease, and Psychosis." In *The Child in His Family: Children at Psychiatric Risk,* ed. E. James Anthony and Cyrille Koupernik. New York: Wiley, 1974.

Cohen, Donald J., et al. "Personality Development in Twins: Competence in the Newborn and Preschool Periods." *Journal of the American Academy of Child Psychiatry* 11:4 (1972), 625–44.

Diamond, Jared M. "Founding Fathers and Mothers." *Natural History*, June 1988.

Dobzhansky, Theodosius. *Mankind Evolving: The Evolution of the Human Species.* 1962. New Haven: Yale University Press, 1975.

Erikson, Erik H. *Childhood and Society*. 2nd ed. New York: Norton, 1963.

———. *Identity: Youth and Crisis.* New York: Norton, 1968.

Farber, Susan L. *Identical Twins Reared Apart: A Reanalysis.* New York: Basic Books, 1981.

Freud, Anna. *The Ego and the Mechanisms of Defense.* Rev. ed. New York: International Universities Press, 1966.

Freud, Sigmund. "Analysis Terminable and Interminable." 1937. In vol. 23 of *The Standard Edition of the Com-*

plete Psychological Works of Sigmund Freud (London: Hogarth Press, 1964).

Frosh, Stephen. *The Politics of Psychoanalysis*. New Haven: Yale University Press, 1987.

Gould, Stephen Jay. *Hen's Teeth and Horses' Toes*. New York: Norton, 1983.

——. *The Mismeasure of Man*. New York: Norton, 1981.

——. "Of Kiwi Eggs and the Liberty Bell." *Natural History*, November 1985, 20–29.

——. *An Urchin in the Storm*. New York: Norton, 1987.

Green, Jonathan, Martin Bax, and Helen Tsitsikas, "Neonatal Behavior and Early Temperament: A Longitudinal Study of the First Six Months of Life." *American Journal of Orthopsychiatry* 59 (January 1989): 82–93.

Hamilton, Annette. *Nature and Nurture: Aboriginal Child-Rearing in North-Central Arnhem Land*. Canberra: Australian Institute of Aboriginal Studies, 1981.

Harre, R. *The Philosophies of Science*. Oxford and New York: Oxford University Press, 1985.

Hartmann, Heinz. *Ego Psychology and the Problem of Adaptation*. New York: International Universities Press, 1958.

Horney, Karen. *New Ways in Psychoanalysis*. New York: Norton, 1939.

Juel-Niesen, N. *Individual and Environment: A Psychiatric-Psychological Investigation of MZ Twins Reared Apart.* 1963. New York: International Universities Press, 1980.

Jung, C. G. "Psychological Types." 1924. In *Readings in Personality,* ed. Harriet N. Mischel and Walter Mischel. New York: Holt, Rinehart and Winston, 1973.

Keller, Evelyn Fox. *Reflections on Gender and Science.* New Haven: Yale University Press, 1985.

Kitchener, Richard F. *Piaget's Theory of Knowledge: Genetic Epistemology and Scientific Reason.* New Haven: Yale University Press, 1986.

Konner, Melvin. *The Tangled Wing: Biological Constraints on the Human Spirit.* New York: Harper Colophon, 1982.

Langer, Jonas. *Theories of Development.* New York: Holt, Rinehart and Winston, 1969.

Lebovici, Serge, and René Diatkine. "Normality as a Concept of Limited Usefulness in the Assessment of Psychiatric Risk." In *The Child in His Family: Children at Psychiatric Risk,* ed. E. James Anthony and Cyrille Koupernik. New York: Wiley, 1974.

Mahler, Margaret. "Rapprochement Subphase of the Separation-Individuation Process." 1972. In *The Process of Child Development,* ed. Peter B. Neubauer. New York: New American Library, 1976.

Mischel, Harriet N., and Walter Mischel, eds. *Readings in Personality*. New York: Holt, Rinehart and Winston, 1973.

Neubauer, Peter B. "Fathers as Single Parents." In *Fathers and Their Families*, ed. Stanley H. Cath, Alan Gurwitt, and Linda Gunsberg. Hillsdale, N.J.: Analytic Press, 1989.

———. "Preoedipal Objects and Object Primacy." *Psychoanalytic Study of the Child* 40 (1985): 163–82.

———. *The Process of Child Development*. New York: New American Library, 1976.

Piaget, Jean. *Play, Dreams and Imitation in Childhood*. New York: Norton, 1962.

Provence, Sally. "Infants in Institutions Revisited." *Bulletin of the National Center for Clinical Infant Programs* 9 (February 1989): 1–4.

Provence, Sally and Rose C. Lipton. *Infants in Institutions*. New York: International Universities Press, 1962.

Rome, Howard P. "Personal Reflections: Personality Disorders." *Psychiatric Annals* 19 (March 1989): 120–21.

Solnit, Albert J. "The Vulnerable Child – In Retrospect." In *The Child in His Family: Vulnerable Children*. ed. E. James Anthony, Cyrille Koupernik, and Colette Chiland. New York: Wiley, 1978.

Sulloway, Frank. *Freud: Biologist of the Mind*. New York: Basic Books, 1979.

Tellegen, A., D. T. Lykken, T. J. Bouchard, K. J. Wilcox, N. L. Segal, and S. Rich, "Personality Similarity in Twins Reared Apart and Together," *Journal of Personality and Social Psychology*, 54 (1988): 1031–39.

Vandenberg, Steven G., et al. *The Heredity of Behavior Disorders in Adults and Children*. New York: Plenum, 1986.

Watson, J. B. *Behaviorism*. New York: Norton, 1924.

Werner, Emmy E. "The Children of the Garden Island." *Scientific American*, April 1989, 106–111.

Wilson, Edward O. *On Human Nature*. New York: Bantam, 1979.

Winnicott, D. W. *The Child, the Family, and the Outside World: Studies in Developmental Relations*. London: Tavistock, 1964.

———. *Through Paediatrics to Psychoanalysis*. London: Hogarth, 1958.

Index

Accommodation, Piaget's theories
  on, 97–98, 173
Adaptation, 78, 93–111, 165
  age and, 101–3
  as avoidance, 103–4
  biological role of, 162
  compensatory behaviors as, 106
  defense mechanisms as, 98–99,
    106–7
  defined, 94
  flexibility and, 95–96 (see also
    Flexibility, human)
  A. Freud on, 98–99, 106–7
  H. Hartmann on, 99
  maladaptive behavior and, 104–
    5
  J. Piaget on, 97–98
  to relocation, 100–1
  vulnerability and, 108–10, 116–
    17
Addictive behavior, 46, 103–4
Adler, Alfred, and nature/nurture
  debate in psychotherapy,
  171–72
Aging, heredity and, 67, 72–73
Alcoholism, 46, 103
Allport, Gordon, 40
Alzheimer's disease, 7, 60, 67, 69
Ambition, trait of, 40–41
Amish population, 68, 185
Anklosing spondylitis, 58–59
Anthony, E. James, 116
Asiatics, The (Prokosch), 164
Assimilation, Piaget's theories on,
  97–98
Autism, 4, 174

Behaviorism, 150–51
Berger, John, 131

Biological demands on psyche,
  176–79
Biological regularity, 157–58
Birth weight, 48–49
Bouchard, Thomas J., 38, 89
Bowlby, John, 3, 127, 174
Brain laterality, 49
Brazelton, T. Berry, 33, 43
Brothers, Leslie, 47

Campbell, Joseph, 16, 177–78
Cancer, heredity and, 67, 156
C. elegans worm, 155–56
Charcot, Jean-Martin, 169
Chess, Stella, 44
Childhood, developmental sequence in, 79–90
Childhood, maturation in, 59–66
  cognitive function in twins, 64–
    65
  disease and, 65–66
  early abilities/faculties in twins,
    63–64
  language ordering, 61
  social smiles (see Smiling, social,
    in infants)
  stranger reaction (see Stranger
    reaction)
  toilet training, 61
Childrearing. See Parenting
Christmas Carol, A (Dickens), 76
Circumcision, infant reaction to,
  117
Cognition
  changes in, 60–61
  characteristics of, 62, 116, 186
  development of, 95, 97
  parenting and risk factors re-

Cognition (*continued*)
lated to, 132, 134–35, 191
in twins, 64–65
Cohen, Donald J., 48, 117
Communes, 192–93
Compulsive behavior
as adaptation, 85, 106
in twins, 20–21
Critical periods, 87, 102
Cultural school theory of psychotherapy, 173

Darwin, Charles, 148
Darwinists, 148–49, 158
Defense mechanisms, 98–99, 106–
7, 116
Deism, 57
Dependency
in childhood, 101–2
in groups, 198
Depression, 46
Development, 2, 75–90, 94–95
discontinuities in, 85–90
individual rates of, 30–34
vs. maturation, 59–60, 77–78
mother-child relationship and,
3–5, 173–74
parental expectations toward
child's, 27–29
sequence of, 79–90
in twins, 80–81, 87–89
Diagnostic and Statistical Manual,
43
Diamond, Jared M., 185–86
Diatkine, Rene, 85
Dickens, Charles, 76, 200
*Discourse on the Origin and
Foundations of Inequalities
Among Men* (Rousseau), 147
Disease in adulthood
environmental factors in, 69–70
heredity, maturation and, 58–
59, 60, 66–68, 144

information about inherited,
transmitted by grandparents,
185–86
in twins, 52, 58–59, 65, 66
DNA (deoxyribonucleic acid), 18,
19, 155–56
apparently functionless sequences of, 159–60
transposability of, 162
Dobzhansky, Theodosius, 64, 79

Egeland, Janice, 68, 127
Ego
development of, 80, 98
functions of, 78, 118, 177–78
hereditary nature of, 107
strength, 105, 107
Ego apparatus, 59, 62, 118
*Ego and the Mechanisms of De-
fense, The* (Anna Freud), 106
*Ego Psychology and the Problem
of Adaptation* (Hartmann), 94
Embryonic development, 161–62
Emotion (affect)
maturation rates and, 70–71
parenting and risk factors related to, 132–34
personality traits related to, 45–
47
Emotional disorders in twins, 87–
89
Empathy, trait of, 2, 47–48, 77
Engageability. *See* Personality
Environment, 8. *See also* Nature/
nurture debate
A. Adler on psychotherapy and
role of, 171–73
adjustment to (*see* Adaptation)
genetic determinism vs., 161–63
impact of, on personality, 45–
51
impact of, on twins, 19–20

influence of, in maturation process, 68–74
interaction of, with heredity, 9–10, 19, 50–53, 201–4
parental role, 3–5
responses in, to individual personality, 27–29, 33–34
risk factors in, 130–37
vulnerability/invulnerability to (*see* Invulnerability; Vulnerability)
Epigenetic guide to life span, 67, 79–90
Erikson, Erik, 33, 67, 79, 80, 163–64, 195, 196
Essay Concerning Human Understanding (Locke), 146–47
Eugenics, 143
Evolution, 48
of adaptability, 95–96, 196
vs. genetic determinism, 158–61
Lamarckian, 150
Experience
constitutional factors in, 169
in early childhood, 170, 174, 177
influence of, 16, 76, 131, 168, 179. *See also* Environment
Extended families, 1, 187–90, 193
Extravert-type personality, 43

Family history, transmission of. *See* Grandparents, role of
Farber, Susan L., 58, 127
Flexibility, human, 95–96, 196. *See also* Adaptation
biological nature of, 100, 162–64
Food preferences in twins, 20, 23
"Founding Fathers and Mothers" (Diamond), 185–86
Frederick II, Holy Roman Emperor, 130

Freud, Anna, 3, 80
on adaptation, 98–99, 106–7
Freud, Sigmund, 78, 95, 107
definition of mental health and conflict by, 84–85
on psychotherapy's limits, 175–76
role in nature/nurture debate in psychotherapy, 168–71, 175–77, 177
Frisch, Max, 117
Fromm, Erich, 173
Frosh, Stephen, 173

Genetic(s), 1, 6–8, 16, 143–44, 153–65. *See also* Heredity
biological survival and, 159–61
chromosomes and DNA, 18, 19, 154–55, 159–60, 162
disease transmission via, 7–8, 60, 65, 66–68, 144
environment and, 161–63
evolution and, 158–61
gene clusters, 156–57
gene proteins, 154
homeobox gene set, 157
personality traits and, 50–52
social behavior and, 163–64
variation in human, 18–19
Genetic clock, 61–62
Genetic determinism, 9–10, 143
environment and, 161–63
evolution and, 158–61
vs. sociobiology, 164
Genetic engineering, 7–8
Goethe, Johann Wolfgang von, 200
Gould, Stephen Jay, 159–60, 179
Grandparents, role of, 1, 183–93
benefits of the role to the grandparents, 189–90
information on inherited disease kept by, 185–86

Grandparents (*continued*)
multiple relationships available
to children through, 187–89
significance of, and family dynamics, 190–93
transmission of family history
by, 184–85
Groups, 195–200
defined, 197
human need for, 196–97
ideology of, 197–99
interdependence of individual
with, 199–200

Hardy, Thomas, 141, 142, 145–46
Hartmann, Heinz, on adaptation,
94, 99, 104
Health, psychological, 103
Freud's definition of, 84
Heart disease, 67, 69
Heredity, 1–3, 15–24. *See also*
Nature/nurture debate
abuses based on differing views
of, 142–43, 149
development and, 82–83
disease and, 66–68
genetics and (*see* Genetic(s))
individual variations in, 30–34
interaction of, with environment, 9–10, 19, 50–53, 201–4
intergenerational transmission of
information about (*see*
Grandparents, role of)
maturation rates based on, 56–59, 60, 61–63
parental expectations and
child's, 27–29
personality traits and, 38–42,
45–47, 50–52
sibling differences in, 16–18, 27
twin's shared, 19–20

vulnerability/invulnerability and
(*see* Invulnerability; Vulnerability)
Hippocrates, 42, 43
Homeobox gene set, 157–58
Horney, Karen, 173
Huntington's disease, 7, 60
Hypnosis, 169

I'm Not Stiller (Frisch), 117
Individuality. *See also* Maturation;
Personality
adaptation and, 96
developmental rates determined
by human, 30–34
groups and, 195–200
heredity-environment interaction
in formation of, 9–10, 19,
50–53
sibling, 16–19
Inhelder, Barbel, 79
In Memorium (Tennyson), 73
Institutionalized care of family
members, 191
Intrauterine influences. *See* Womb
environment
Introjection, 98
Introvert-type personality, 43
Invulnerability, 124–28
adaptability and, 116–17
characteristics of, 115–16
defined, 125
dysfunctional families, poverty,
and, 126–28
psychological problems and,
114–15

Jefferson, Thomas, 57, 148
*Journey Not the Arrival Matters,
The* (Woolf), 200
Jung, Carl, 43, 173

Keller, Evelyn Fox, 162
Kibbutzim in Israel, families on, 192–93
Kiwi bird, 160
Konner, Melvin, 157

Language development, 61, 71, 130
Learning, 4
  and bowel control, 61
  empathy and, 47
  image of, 77
  in opposition to biology, 57, 145, 151, 172
  from parents, 21, 78, 102
  as partner of biology, 8, 11, 77
  problems (see Learning problems)
Learning problems, 30, 102, 116, 126
  dyslexia, 106
Learning theory, Skinner's, 150–51
Lebovici, Serge, 85
Lipton, Rose C., 191
Locke, John, 146–47, 148

McClintock, Barbara, 162
Mahler, Margaret, 3, 71, 80, 174
Maladaptive behavior, 104–5
Manic-depressive disorder, 7, 46, 68
Maturation, 2, 55–74, 94–95
  in childhood, 59–66
  vs. development, 59–60, 77–78
  disease in adulthood and, 66–68
  genetic basis of, 56–59, 60
  influence of environment in, 68–74
Minnesota Multiphasic Personality Inventory, 43

Motor ability, 59, 60, 64–65, 134. See also Sensorimotor ability
  maturation rate of, 70–71

Natural selection, 158–59. See also Evolution
Nature/nurture debate, 2–3, 8–9, 141–52, 201–4. See also Environment; Heredity
  Locke's and Rousseau's environment arguments, 146–48
  popular views of heredity in, 142–46
  psychotherapy and, 167–79
  Skinner's behaviorism, 150–51
  social abuses rooted in, 142–43, 149–50
  social Darwinists view in, 148–49, 158
Neonatal Behavior Assessment Scale, 43
Neurosis
  adaptation and, 99
  development and, 84–90
Normality/abnormality issues in development, 85–90
Nutrition, 68–69

Object relations theory in psychotherapy, 173–74
Oncogene, 67
Origin of the Species, The (Darwin), 148

Parenting
  appropriate nurturing for the individual child, 81–84, 132–37
  mother-child interaction, 3–5, 170, 173–74
  parental adaptations, 33–34
  parental attitudes, 20, 21
  parental affective state linked to emotion in child, 45–47

Parenting (*continued*)
parental expectations, 27–29
permissiveness in, 135–36
risk factors in, 132–37
single-, 99
of vulnerable children, 121–23
Personality, 37–53. *See also* Individuality
development and, 80
engager-type, 17–18, 39, 42, 44–45, 116, 121
environmental impacts on, 49–51
environmental reactions to, 27–29, 33–34
extravert-type, 43
family history and traits of, 186
hereditary links to traits of, 38–42, 45–47
identifying basic traits of, 42–48
interaction of environment and heredity in, 50–53
intravert-type, 43
responder-type, 16–17, 45
sibling differences in, 16–19
time factors in, 39 (*see also* Maturation)
womb environment and, 48–49
Phobias, 89, 105–6
Piaget, Jean, 3, 79, 95
on adaptation, 97–98
Popper, Karl, 40
Poverty, impact of, on individuals, 126–27, 130–31
Projection, 98–99
Prokosch, Frederic, 164
Provence, Sally, 191
Psychological problems
emotional disorders in twins, 87–89
heredity and psychosis, 67–68
sleep disorders, 123–24

vulnerability to paranoid schizophrenia, 114
Psychosexual stages of development, 79–90
Psychotherapy, nature/nurture debate in, 167–79
A. Adler's work on, 171–73
biological demands on psyche and, 176–79
cultural school and, 173
ego functions and, 177–78
S. Freud's work on, 168–71, 175–77
limits of psychotherapy and, 175–76
object relations theory and, 173–74
Puberty, 66–67, 68–69

Racism, 3, 143, 149, 152, 164
Reactive capabilities, 116–117
Relocation, adaptation to, 100–101
Resilience. *See* Invulnerability
Responder-type personality, 16–17, 45
Rilke, Rainer Maria, 29–30
Risk factors in the environment, 129–37
Rousseau, Jean-Jacques, 131, 146, 147, 148

Schizophrenia, 68, 114, 174
Sensorimotor ability, 62, 97, 116, 177, 186
Separation problems, 123
Shakespeare, William, 42
Shyness, 1, 11
as compensation, 106, 135
as fundamental characteristic, 38, 39, 40, 47, 50, 51
Sickle-cell anemia, 60

Single-parent families, 99, 130–31, 184–93
Skinner, B. F., 150–51
Sleep disorders, 123–24
Smiling, social, in infants, 60, 133
Sociobiology, 164
Solnit, Albert J., 126
Spock, Benjamin, 151–52
Stranger reaction in infants, 61, 121, 133
Stravinsky, Igor, 35
Sulloway, Frank, 170–71
Survival and genetics, 159–61
Susceptibility. *See* Vulnerability

Tabula rasa concept, 127, 147
Tennyson, Alfred, 73, 199
Tess of the D'Urbervilles (Hardy), 141, 142, 145–46
Thalidomide, 161
Toilet training, 61
Tradition, 34–35, 184, 192, 204
Trait clusters, 44
Trauma, effect of, 4, 170, 177
Twin(s)
  addictive behavior in, 102–4
  cognitive function in, 64–65
  compulsive behavior in, 20–21
  development studies of, 80–81, 87–89
  disease in, 52, 58–59, 65, 66
  effect of birth weight and womb environment on, 48–49
  emotional disorders in, 87–89
  environment's impact on, 19–20
  food preferences in, 20, 23
  identifying personality traits in, 44–48
  maturation in, 58–59, 63–66
  retrospective vs. prospective studies on, 5–6
  shared behavior traits in, 20–23
  shared genetics in, 19
  vulnerability in, 120–22
Twinning reaction, 23
Type A behavior in infants, 39

Vulnerability, 115–24
  adaptability and, 116–17
  in an adult, 118–20
  characteristics of, 115–16
  environmental risk factors and, 129–37
  psychological problems and, 114–15
  sensitivity levels and, 116, 118, 120, 121
  in twins, 120–22

Watson, J. B., 150–51
Winnicott, D. W., 29, 34, 174
Wizard of Oz, The (movie), 104
Womb environment, 48–49
Woolf, Leonard, 200
Wundt, Wilhelm, 42–43